THE NEW REGIONAL ECONOMIES

Cities & Planning Series

The *Cities & Planning Series* is designed to provide essential information and skills to students and practitioners involved in planning and public policy. We hope the series will encourage dialogue among professionals and academics on key urban planning and policy issues. Topics to be explored in the series may include growth management, economic development, housing, budgeting and finance for planners, environmental planning, GIS, small-town planning, community development, and community design.

Series Editors

Advisory Board of Editors

William R. Barnes
Larry C. Ledebur

THE NEW REGIONAL ECONOMIES

The U.S. Common Market
and the Global Economy

UPA LIBRARY

Cities & Planning

SAGE Publications
International Educational and Professional Publisher
Thousand Oaks   London   New Delhi

For information:

 SAGE Publications, Inc.
2455 Teller Road
Thousand Oaks, California 91320
E-mail: order@sagepub.com

SAGE Publications Ltd.
6 Bonhill Street
London EC2A 4PU
United Kingdom

SAGE Publications India Pvt. Ltd.
M-32 Market
Greater Kailash I
New Delhi 110 048 India

Printed in the United States of America

Library of Congress Cataloging-in-Publication Data

Barnes, William R.
   The new regional economies: the U.S. common market and the global
economy / by William R. Barnes and Larry C. Ledebur.
      p. cm. — (Cities and planning; v 2)
   Includes bibliographical references and index
   ISBN 0-7619-0938-9 (Cloth: acid-free paper). — ISBN 0-7619-0939-7 (pbk.:
acid-free paper)
      1. United States—Economic conditions—1981- —Regional
disparities. 2. Intergovernmental fiscal relations—United States.
3. Regional economics. 4. Competition, International. I. Ledebur,
Larry C. II. Title. III. Series.
HC106.8.B365 1997
330.973—dc21   97-4751

This book is printed on acid-free paper.

98 99 00 01 02 03 10 9 8 7 6 5 4 3 2 1

Acquiring Editor:	Catherine Rossbach
Editorial Assistant:	Kathleen Derby
Production Editor:	Sherrise M. Purdum
Production Assistant:	Denise Santoyo
Typesetter/Designer:	Janelle LeMaster
Indexer:	Teri Greenberg
Print Buyer:	Anna Chin

For *Éva and Catherine*
　　　　　—WRB

❧⁓❧

For *Susan, Kathryn, Lara, and Elisa*
　　　　　—LCL

CONTENTS

SERIES EDITORS' INTRODUCTION

The study of cities is a dynamic, multifaceted area of inquiry which combines a number of disciplines, perspectives, time periods, and numerous actors. Urbanists alternate between examining one issue through the eyes of a single discipline and looking at the same issue through the lens of a number of disciplines to arrive at a holistic view of cities and urban issues. The books in this series look at cities from a multidisciplinary perspective, affording students and practitioners a better understanding of the multiplicity of issues facing planning and cities, and of emerging policies and techniques aimed at addressing those issues. The series focuses on both traditional planning topics such as economic development, management and control of growth, and geographic information systems. It also includes broader treatments of conceptual issues embedded in urban policy and planning theory.

The impetus for the *Cities & Planning* series originates in our reaction to a common recurring event—the ritual selection of course textbooks. Although we all routinely select textbooks for our classes, many of us are never completely satisfied with the offerings. Our dissatisfaction stems from the fact that most books are written for either an academic or practitioner audience. Moreover, on occasion, it appears as if this gap continues to widen. We wanted to develop a multidisciplinary series of manuscripts that would bridge the gap between academia and professional practice. The books are designed to provide valuable information to students/instructors and to practitioners by going beyond the narrow confines of traditional disciplinary boundaries to offer new insights into the urban field.

Bill Barnes and Larry Ledebur's *The New Regional Economies* represents a new and thought provoking view of regional economies and regional problem-solving and governance. Their book offers a natural evolution of their earlier, and often cited, work on cities and the national economy. *The New Regional Economies* describes how local economies transcend local boundaries and need to be viewed in a new context. We are excited about the book's potential to elevate the importance of local economies in the national economic context and to place the discussion on the national policy agenda. This well-written and well-documented book will generate a robust discussion among policy makers and academics in both theory and practice.

—Roger W. Caves,
San Diego State University

—Robert J. Waste,
California State University at Chico

—Margaret Wilder,
State University of New York at Albany

FOREWORD

Donald J. Borut,
Executive Director
National League of Cities

The ideas and research advanced by Bill Barnes and Larry Ledebur in *The New Regional Economies* constitute a major shift in thinking about cities and economy. These ideas make a big difference for city governments and also for state and federal officials and for those who care about cities and about economics. Local economic affairs and policy, rather than abstract national averages, become the center of attention.

Their work has already provoked and made contributions to several fields of study and public discussion. Their vocabulary has become state-of-the-art talk—local economic regions, local economies, the U.S. Common Market, and now the Regional and Global Economic Commons.

By providing an economic rationale for regional problem-solving and governance, they contribute a new, more constructive basis for the often controversial issues around city/suburbs relationships and the broader discussions about regionalism. Their focus on governance rather than the structural fix of metro government supports a general shift in that direction. By formulating and explaining the idea of a "local economic region," their work creates a framework, beyond jurisdiction-bounded thinking, for local economic development. Their metaphor of the "U.S. Common Market" highlights the importance of local economies in the national economic context.

Bill and Larry's collaboration has been creative and productive. It began as annual economic reports to the NLC Board of Directors. They then articulated a direction for further research and policy analysis that served as the basis for a successful proposal to the Ford Foundation for a project "focused on cities and the national economy." They first publicly described the outlines of their framework in a presentation at the 1990 annual conference of the Urban Affairs Association and then in "Toward a New Political Economy of Metropolitan Regions" in *Government and Policy*, 1991. That framework also laid the foundation for a series of widely-reported studies of city-suburbs economic relations from the National League of Cities: *City Distress, Metropolitan Disparities, and Economic Growth; All in it Together: Cities, Suburbs, and Local Economic Regions;* and *Local Economies: The U.S. Common Market of Local Economic Regions.*

At NLC, we have combined this line of thinking with evolving local practice into an agenda of other NLC activities: revisions of NLC's Federal policy statements, several of our annual "Futures" reports, some technical assistance activities and reports. In 1996, under the leadership of NLC President Gregory Lashutka, Mayor of Columbus, Ohio, we convened a national Symposium on "Achieving World Class Local Economies" and, based on that exciting conference, we produced a guidebook for city officials by the same title. We plan to continue to develop this agenda of activities.

The ideas that Bill and Larry have developed are a prominent part of an evolving body of research and public discourse on this nexus of topics. This book places them again at the forefront of this thinking. They create a large, systematic framework that moves the city/suburbs/regionalism issues out of the narrow box of local government structure and into the larger context of economic performance and economic policy. This transformation helps us all to get past some old deadends of urban policy and, indeed, to put the whole of urban policy in a new light. Perhaps even more important, *The New Regional Economies* points toward a rethinking of economic policy issues with local realities at the center of attention.

The work on this book was supported by NLC in its research agenda and by a 1988 grant by the Ford Foundation to the National League of Cities Institute, NLC's educational and research affiliate. On behalf of NLC and the leaders of the nation's cities, I am pleased to say that this is work of which we are rightly proud.

ACKNOWLEDGMENTS

Our journey to this book has been long, exciting, and congenial. We enjoyed doing it together.

We thank the many people who helped, supported, and/or provoked us along the way.

NLC has been an engaged and supportive environment. Executive Director Donald J. Borut and Deputy Executive Director Christine Becker were patient; they also prodded and helped use our thinking toward practical outcomes for cities. Previous NLC executives Alan Beals and William Davis saw the potential of the ideas and the project and supported the undertaking. Many NLC staff and member city officials—especially Mayor Gregory Lashutka of Columbus, Ohio—stimulated our work with enthusiasms and pertinent questions and by making use of the ideas.

The Ford Foundation provided a research grant in 1988 to the National League of Cities Institute to help support this project. Program Officers David Arnold and Michael Lipsky tolerated many delays.

Larry Ledebur's contributions to this book occurred in the conducive settings of three universities: the University of South Carolina, Wayne State University, and Cleveland State University.

We presented our ideas at several stages of this work; the ensuing discussions helped us evolve our thinking. The presentations included panel sessions at the annual conference of the Urban Affairs Association on three separate occasions; formal reports to the NLC Board of Directors, the NLC Executive staff, and the annual meeting of the Executive Directors of state municipal leagues; two day-long seminars that we convened in 1989 and again in 1990; the 1994 symposium, "The New Regionalism," put on by the Social Science Research Council for the U.S. Department of Housing and Urban Development; and the 1996 symposium of the North American Institute for Comparative Urban Research.

Student interns contributed research support at various times: Britany Orlebeke, Kevin Khayat, David Dickinson, Amy Bell, Kyla Weisman.

Bill Woodwell edited final versions of chapters and consolidated them into a single manuscript.

We also thank the following friends and colleagues for their contributions: Mike Garn, Niles Hansen, Hugh Knox, Dave Garrison, Arthur Nelson, Susan Clarke, Michael Pagano, Royce Hanson, Harold Wolman, John Ross, Ned Hill, Jeff Henig, Frances Frisken, Todd Swanstrom, Don Phares, Dennis Judd, Allen Artibise, David Rusk, Neal Peirce, Marc Weiss, and Bob Waste.

Bill Barnes and Larry Ledebur

1

INTRODUCTION
AND OVERVIEW

The pervasive image of a single "U.S. economy" contained within the nation's geographic borders is the lens that currently focuses public policy in the United States. It provides the rationale for the prevailing national orientation of economic policies.

Political jurisdictions, however, are polities, not economies. These include nations, states, and local governmental jurisdictions. The implications of this simple observation are complex and far-reaching.

Local economies—primarily metropolitan-centered and strongly linked—are the real economies in the United States. The boundaries of the nation create within them a national common market of these local economies, most of which are also linked to economies abroad.

Therefore, the conventional view of the nation as an economic unit without important internal differentiation must be put aside. It is not correct. Macroeconomics is enthralled with this view, just as public discussion is trapped by it. All of us are thus harmed. If U.S. economic performance is to be improved and if Americans are to restore their

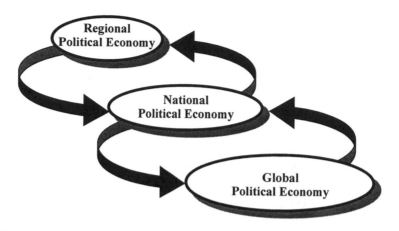

Figure 1.1. Interdependent Economic Systems

sense of confidence in government, then a better framework is required. Policymakers will have to dethrone their fascination with the nationalist view of the economy and come to grips with local and global realities.

A three-tiered political economy—local, national, and global—is the appropriate framework for economic thinking and policy making. Figure 1.1 presents an image of this framework. The key economic relationship is among local or "regional" economies. To understand and be able to affect this relationship, however, we must see beyond the local situation to the national and the global. Conversely, we must be able to bring "national" and "global" issues to ground. No part of the system is autonomous. Each is linked to—and, indeed, reflected in—the other.

We need a way of seeing economics that highlights these basic patterns. The nationalist viewpoint shrouds them.

In *The Work of Nations,* Robert Reich aptly defined the national economy as "the region of the global economy denominated by the nation's political borders."[1] Within those borders, author Jane Jacobs observed, "most nations are composed of collections or grab bags of very different economies."[2] Workers, firms, governments, and households all do what they do within this spatial framework. This applies to a multilocation corporation as well as to a displaced worker migrating to a new job.

We do not argue that no U.S. economy exists. The powers of a strong central government, along with shared history and culture, have created a national economic entity. Our argument is that the "national economy" is spatially differentiated and that local economic regions are the crucial units for focusing analysis and policy. The national economy is a common market of these regional economies.

As a result, uniform economic policies that respond to national average indicators rather than local conditions will not maximize economic efficiency, productivity, and growth. Rather, the effectiveness of federal, state, and local economic policies depends, in part, on each government's ability to correctly identify the structure and dynamics of the U.S. economy. Similarly, the accountability of public officials is partially dependent on citizens having an understanding of the basic economic construct underlying economic policy decisions. Thus, policy and politics, efficacy and democracy, rely to a large extent on an adequate concept of what the U.S. economy is and how it works.

An inappropriate economic framework leads people to perceive and describe problems incorrectly, and this inevitably leads them to select solutions that do not work. This is what is wrong with the current one-size-fits-all, "rising tide lifts all boats," nationalist economic framework. It is time for it to go.

The Federal Reserve, for example, defines its inflation and business cycle concerns and applies its policy tools according to measures that summarize conditions in the whole nation. A national average change in prices of $X\%$ will mask a potentially wide range of changes in the local economies. A $Y\%$ change in the discount rate is the same all over, however, producing too little effect in economies with high price change and too much effect in places with low price change. Although there may be a few local economies for which the national average policy is appropriate—and although the national average indicator may show the desired overall effect—the functional local economies will bear various and mostly inappropriate impacts.

The mesmerizing power of the "economy as nation" perspective also hurts American politics and democracy. The nationalist view sets up false choices between local and national interests, grouping the former with other "special interests" and treating the latter as somehow the sole locus of the general good. In addition, this perspective allows some local public officials and citizens to treat "the economy" as something beyond their control, as an external "force" that affects their communities like the weather, bad or good. It also encourages some federal and state officials to lay claim to and use powers over the economy that ignore the variety of local conditions and end up harming local communities in the process. The nationalist perspective thus befuddles accountability and weakens democracy.

Shifting away from the prevailing economic view is especially important for citizens and officials of local governments. Such a shift provides a rationale and a framework for their participation in national economic policy making on behalf of their local economies. Also, because local economies are not defined by the boundaries of any local governmental jurisdiction, some forms of interlocal governance are necessary to establish legitimate standing for local officials in economic policy making. Although no easy matter, relinquishing some authority may be the gambit for obtaining larger power. Furthermore, some citizens and officials will have to give up the fiction that the government's responsibility is only for "delivering services." They will have to surrender a jurisdiction-based view of the local economy and focus instead on the underlying areawide economy and its connections to the world all around.

The regional economic framework that we advance in these pages also has important consequences for state and especially federal policy. States should empower and encourage interlocal arrangements for economic policy making and target their programs accordingly. Similarly, federal fiscal and monetary policies, as well as tax and spending programs, should focus on enhancing the performance of the local economic regions.

In this book, we build the case for a new, more realistic, and more useful way of thinking about the U.S. economy. Then, we suggest what differences a regional economic framework will have for policy, policy making, and how government works. Seen through a different lens, important current issues—proposals for "sorting out" government responsibilities, the idea of "urban policy," and the criteria for assessing international trade agreements—take on fresh aspects. Interestingly, some issues that are now cast as either/or choices can be reframed toward syntheses and more constructive outcomes.

Thus, for example, it is fruitless to debate whether suburbs are "independent" of central cities or whether central cities are the irreplaceable "engines" for metropolitan economies. In a local economy, city and suburb are interdependent parts of the whole; their roles and relationships need to be specified in each locale. Similarly, the metropolitan government versus fragmentation debate can and should give way to a more useful synthesis for options for governance and policy making that can nurture the regional economic commons. Subsequent chapters will explore these and other dead-end dichotomies—national-local, local-global, and economics-politics—that might be reformulated or synthesized to be more useful in light of the regional framework.

In recent decades, economic policy discussion has followed two tracks. Each contains liberal and conservative elements. These tracks are distinguished not by political viewpoint but by their underlying frameworks. The dominant view is purely nationalist, defining the U.S. economy according to the national averages churned out by our inadequate data system. This view also treats trade and other international economic relations in strictly nationalist terms; that is, using U.S. boundaries to distinguish between "our" economy and others.

At the same time, a widening stream of thought focuses on significant components of economic activity within U.S. borders and on how changes in global relations should force us to rethink the American economy.

When the whole of this second stream is considered, it constitutes more than a simple revision of the purely nationalist framework. Although their shared underlying perspective is often obscured by the policy debates in which proponents make their case, these countervailing views distend the older framework beyond recognition.

The regional framework we offer carries that second stream of thought to new terrain, where it becomes clear that the U.S. economy is in reality a common market of local economies, most of them centered on metropolitan areas.

Because its focus shifts from abstract national data to tangible local realities and linkages among the local economies, the regional framework calls on government entities at all levels to adjust their policies to address the needs of these local economies. Also, because no government is congruent with any real economy, governments will need to collaborate. The tempting simplicity of "sorting out" functions intergovernmentally must be forsworn.

Perhaps because the conventional approach is fully statist, the second stream has been vulnerable to the opposite error of economism. At the extremes, the first defines reality in terms of governmental jurisdictions; the other sees only free-floating firms and workers. This thesis-and-antithesis posture of the two perspectives has the appeal of simplicity and ease of debate. It does not, however, much help citizens and public officials who are trying to address real problems that involve real people in real places.

The relevant question is not whether the world is chiefly an economic or a political construct; the question is how do polities and economies exist together to create the political economies we all live in and seek to improve.

There is no "free market" and no government program that will automatically (or even over the long run) solve our economic problems for us. Thus, we are obliged to seek a better understanding of the political economy to address our problems more successfully. The chapters that follow take the idea of political economy apart and put it together again to create a better framework for understanding and problem solving. We have not tried to add a full social and cultural dimension to this framework, although such an addition will be important.

This book, then, is about ideas. Also, it is about doing something—specifically, making economic policy—as a result of the ideas we propose. In this, we follow John Maynard Keynes and Richard Weaver. From Keynes:

> The ideas of economists and political philosophers, both when they are right and when they are wrong, are more powerful than is commonly understood. Indeed the world is ruled by little else . . . I am sure that the power of vested interests is vastly exaggerated compared with the gradual encroachment of ideas.[3]

And from Weaver:

> *Ideas Have Consequences.*[4]

Moving from the conventional, nationalist view of the economy to the regional image proposed here would be a "paradigm shift." The latter concept comes from the work of the late Thomas S. Kuhn in *The Structure of Scientific Revolutions.*[5] Kuhn confined his thesis to science only, but it has entered the popular vocabulary and been widely misused. Despite the risks, we think his description is useful.

Kuhn said that paradigm shifts, or revolutions, are the means by which transforming perspectives are incorporated into conventional science. They occur when the current paradigm cannot account for increasing anomalies—evidence and conditions that cannot readily be explained. These anomalies, in turn, warrant the creation of a new framework.

Kuhn described the paradigm shift as a process of "discovery." Discovery commences with the awareness of anomaly—that is, with the recognition that events and conditions have somehow violated the paradigm-induced expectations that govern normal science. It then continues with a more or less extended exploration of the area of anomaly. It closes only when the paradigm has been changed so that the anomalous has become the expected.

This book roughly follows this process. Chapter 2 provides a brief history of the rise and triumph of the nationalist paradigm. Chapters 3, 4, and 5 describe anomalies, evidence, and conditions that cannot adequately be explained by this paradigm. These discussions lay the groundwork for the presentation of a quite different way of looking at the economy and, consequently, at the political economy (Chapters 6-8). The ambition of these three chapters is to present an alternative paradigm, one that will offer greater explanatory power and, therefore, be a more effective guide for decision making.

Chapters 9, 10, and 11 begin the application of our alternative paradigm for public policy and governance. The significance of the regional economic paradigm is that it shapes the way people see the world and, thus, determines and defines the opportunities and the challenges they will be able to perceive. In Chapter 12, we review the terrain we have covered and identify some areas for new explorations.

We do not claim that this essay completes the paradigm shift whose necessity we proclaim. We do hope, however, that it makes a substantive contribution to this end.

NOTES

1. Reich, R. (1991). *The work of nations: Preparing ourselves for the 21st century capitalism* (p. 244). New York: Knopf.

2. Jacobs, J. (1984). *Cities and the wealth of nations: Principles of economic life* (p. 31). New York: Random House.

3. Keynes, J. M. (1964). *The general theory of employment, interest, money* (p. 384). New York: Harcourt, Brace, Jovanovich. (Original work published 1936).

4. Weaver, R. M. (1948). *Ideas have consequences.* Chicago: University of Chicago Press.

5. Kuhn, T. S. (1970). *The structure of scientific revolutions* (2nd ed., pp. 52-53). Chicago: University of Chicago Press.

THE NATION AS ECONOMY
Triumph of a Faulty Paradigm

The economist must become a student of political science
in order to answer strictly economic questions.

—Scott Greer, *The Emerging City: Myth and Reality*[1]

Most current economic thinking is rooted in the concept that the nation is the operative unit of economic analysis. The origins of this idea are deeply embedded in the historical struggle for ascendancy between city-states and nations as well as in the promise of Keynesian economics that the capitalist demon of the business cycle, recurrent swings of prosperity and wrenching depressions, can be controlled through nationally administered macroeconomic policy.

HISTORICAL PERSPECTIVE:
THE RISE AND DEMISE OF THE CITY-STATE

In the historical struggle for supremacy, nation-states—the centers of power—gained ascendancy over city-states—the centers of wealth. The

French historian Fernand Braudel, in *The Perspective of the World*,[1] undertakes to trace and explain the rise of the city-centered economies of the European past and their eventual subjugation and domination by nation-states.

According to Braudel,[2] the age of empire-building cities began with the Hansa guilds of Germany and ended with Amsterdam, the last time "a veritable empire of trade and credit could be held by a city in her own right, unsustained by the forces of the modern states." The burning question in medieval Europe was which of the potential enemies, cities or states, was to dominate the other.

At their zenith, city-states ruled world trade. These city economies were modern forces that took advantage of the backwardness and inferiority of other areas to expand, to dominate, and to exercise a near-monopoly power over the large profits of long-distance trade.[3] Braudel argues that

> There is nothing very surprising about this: Towns, as all historians agree, have been the essential instruments of accumulation, the motors of economic growth, the forces responsible for all progressive division of labour. . . . Merchant capitalism, by circumventing the restrictive practices of the urban guilds, thus created a new industrial arena—in the countryside but controlled from the towns. For everything came from the towns, everything started there.[4]

Somehow, these city economies—centers of world commerce and political intrigue—lost their struggle for supremacy with nation-states. Braudel says that four factors, or forces, appear to have caused the nation-state's rise to dominance.

First, the city-state, with few resources within itself, constituted a very narrow foundation from which to dominate neighboring economies and world markets. By the end of the eighteenth century, Braudel believes, with the growth of the world economy and the beginnings of the Industrial Revolution in Europe, this base became too narrow to maintain its hold on the structure of trade and commerce. Independently, cities lacked sufficient control over the neighboring economies on which their success depended. As the city-states' span of control eroded, the nation-state became ascendant.

The second factor contributing to the demise of the city-state was the primacy of politics—the ascendancy of politics over economics. Nation-states—political entities—were not economic units but constructs organized according to political boundaries. From Braudel's perspective,

> An economic area always extends far beyond the borders of political areas. Nations or national markets were consequently built up inside an economic system greater than themselves, or more precisely they were formed in opposition to that system. A long-range international economy already existed and the national market was carved out within this wider unit by more or less far-sighted and certainly resolute policies. Well before the age of mercantilism, princes were already intervening in the economy, seeking to constrain, encourage, forbid, or facilitate movement, filling a gap here or opening an outlet there. They sought to establish regular systems which would assist their own survival and political ambitions, but they were only successful in this endeavor in the end if it coincided with the general tendencies of the economy.[5]

A third factor leading to the eventual triumph of the nation-state was the rise of mercantilism, a force closely related to the ascendancy of politics over economics. Braudel stated that "mercantilism represents precisely the dawning of awareness of this possibility of maneuvering the entire economy of a country—in fact, it could be described as the first attempt to create a national market."[6]

Indeed, mercantilism has been defined as the transfer of the control of economic activity from the local community to the state.[7]

Finally, it may be that economic growth was the truly decisive factor in the expansion and consolidation of national markets and the resulting disjuncture of the locus of economic wealth and political power. According to Braudel,

> A gulf developed between nation-states on the one hand, the locus of power, and urban centers on the other, the locus of wealth. . . . It took the economic miracle of the eighteenth century to remove the last obstacle, leaving the economy from now on under the aegis of states and their national markets, the heavyweights to whom the future belonged. It is not surprising then that the territorial states, though having tasted political success early on, should have come late to the economic success represented by the national market, the promise of their material triumphs.[8]

The ascendancy of politics over economics in the development of the nation-state is more than a footnote of interest only to students of history. The origins of contemporary economic thought and theory can be traced to the rise of mercantilism and the centralization of power in the nation-state. Indeed, the noted historian of economic thought, Sir Alexander Gray, believed that mercantilism was never more than a means to a political end: the creation of a strong state.[9]

COMPLICITY OF ECONOMICS AND THE DEMISE OF POLITICAL ECONOMY

The emergence of the nation-state created what is now referred to as the national economy by imposing political boundaries on emerging patterns of market organization and development. A central thrust of economics and economic policy in the past century has been the effort to bring rationality and coherence to this artificial economic construct. Under nationalism, the boundaries of the economy and the boundaries of the polity, the nation, became one and the same. The nation and the national economy were one.

Having lost the battle of boundaries to politics, economics has gone on to win the war for control of national values and decision making.[10] Political culture in the United States is to a large degree defined by the fact that the "economy" and the nationalist paradigm have gained ascendancy. In the age of nationalistic mercantilism, politics were ascendant over economic interests. In the nineteenth century, however, as the forces of modern capitalism displaced mercantilism, "economic activity became a system of market relationships freed of all constraint, practical and moral, beyond itself."[11] Also, by the end of World War II, economics had assumed the status of a "science" and the "market" had become an arbiter, an impartial decision-making entity that required no guidance from the visible hand of politics except to "tip the scales."[12]

CHALLENGES TO NATIONALISM

The nationalist paradigm and its statist view of the "nation as economy," although dominant, is not unchallenged. Jane Jacobs, writing in the early 1980s, argued that cities, not nations, are economies and the economic engines of growth. Recently, Kenichi Ohmae, addressing the concept of a global economy without national borders, shined new light on the diminishing economic roles of nation-states. For both Ohmae and Jacobs, the underlying thesis is that nations are not the real economies. Both also embrace an economism in which economics transcends politics.

In *Cities and the Wealth of Nations: Principles of Economic Life*[13] Jacobs, echoing Braudel, argued that cities are the true source of energy in national economic life. Although this provocative book attracted wide attention, the economics profession, always alert to incursions by the unlicensed, appeared unmoved by this challenge to contemporary economic thought.

Jacobs launched a frontal attack on the "failed promise of economics." She wrote, "For a while in the middle of this century, it seemed that the wild, intractable, dismal science of economics had yielded up something we all want: instructions for getting and keeping prosperity."[14] She concluded, however, that this hope is a "fool's paradise." She continued:

> Macroeconomics—large-scale economics—is the branch of learning entrusted with the theory and practice of understanding and fostering national and international economies. It is a shambles. Its undoing was the good fortune of having been believed in and acted upon in a big way—never has a science, or supposed science, been so generously indulged. And never have experiments left in their wakes more wreckage, unpleasant surprises, blasted hopes and confusion, to the point that the question seriously arises whether the wreckage is repairable; if it is, certainly not with more of the same.[15]

Jacobs urged economists to go "back to reality." She argued that the economist's habitual practice of simply trying to use existing tools with "greater sophistication, shuffling the same old conceptions into new combinations and permutations" will not work because a basic assumption, taken for granted in the mystique of economics, is in error:

> Macro-economic theory does contain such an assumption. It is the idea that national economies are useful and salient entities for understanding how economic life works and what its structure may be: that national economies and not some other entity provide the fundamental data for macro-economic analysis.[16]

Surely, Jacobs maintained, no other body of scholars or scientists in the modern world "has remained as credulous as economists, for so long a time, about the merit of their subject matter's most formative and venerable assumption." Jacobs stated,

> Nations are political and military entities, and so are blocs of nations. But it doesn't necessarily follow from this that they are also the basic, salient entities of economic life or that they are particularly useful for probing the mysteries of economic structure, the reason for the rise and decline of wealth. Indeed, the failure of national governments and blocs of nations to force economic life to do their bidding suggests some sort of essential irrelevance.[17]

The fallacious assumption that nations are the salient economic entities is more than four centuries old, according to Jacobs, and its source is in nationalistic mercantilism. Early mercantilists, she argued, assumed that national rivalries were the keys to understanding wealth and its creation. Under this naive mercantilism, wealth was equated with precious metals, and the key to national accumulation of wealth was to export more than was imported. Defining wealth as national accumulation of specie automatically made nations the salient economic entities.[18]

If we remove the "blinders of mercantilist tautology," according to Jacobs, we see the economic world in its own right rather than as an artifact of politics.[19] Nations are not economies but rather grab bags of diverse economies, and among these less-than-national economies, cities are unique in their abilities to shape and reshape the economies of other settlements, including those far removed from them geographically.[20]

Jacobs' book is part of a line of research that lies, by and large, outside the mainstream of contemporary economic theory. It has been largely ignored by an economics profession deeply mired in the nationalist economic paradigm. A recent challenge to the nationalist economic paradigm is articulated in Ohmae's two books, *The Borderless World: Power and Strategy in the Interlinked Economy* and *The End of the Nation State: The Rise of Regional Economies*. Two central themes interweave in Ohmae's argument: the diminished economic role of national governments and the emergence of natural economic zones or region-states.

Central to Ohmae's thesis of the decline of the nation-state is the role of technology and technological breakthroughs, especially in communications and finance:

> On a political map, the boundaries between countries are as clear as ever. But on a competitive map, a map showing the real flows of financial and industrial activity, those boundaries have largely disappeared. Of all the forces eating them away, perhaps the most persistent is the flow of information—information that governments previously monopolized, carving it up as they saw fit and redistributing it in forms of their own devising.[21]

The spinning out of these forces, Ohmae argues, raises troubling questions about both the relevance and the effectiveness of viewing national governments as the construct for thinking about—and, much less, managing—economic activities.[22] Future economic prosperity, according to Ohmae, requires limiting the power and role of nation-states and granting increasing autonomy to wealth-producing regions. Ohmae stated,

> In today's borderless economy . . . there is really only one strategic degree of freedom that central governments have to counteract this remorseless buildup of economic cholesterol, only one legitimate instrument of policy to restore sustainable and self-enforcing vitality, only one practical as well as morally acceptable way to meet their people's near-term needs without mortgaging the long-term prospects of their children and grandchildren. And that is to cede meaningful operational autonomy to the wealth-generating region states that lie within or across their borders, to catalyze the efforts of those region states to seek out global solutions, and to harness their distinctive ability to put global logic first and to function as ports of entry to the global economy. The only hope is to reverse the postfeudal, centralizing tendencies of the modern era and allow—or better, encourage—the economic pendulum to swing away from nations and back toward regions.[23]

Until this happens, according to Ohmae, we are captives of a "cartographic illusion" in which public debate is held hostage to the outdated language of political borders while citizens and consumers speak in the vastly different idiom of the global marketplace.[24]

City-regions or region-states are the true sources of economic energy in the global economy, Ohmae argues, not the nations or nation-states of the nationalist paradigm.[25] Examples identified by Ohmae include northern Italy, San Diego-Tijuana, Hong Kong-Southern China, California's Silicon Valley-Bay Area, the Research Triangle of North Carolina, and so on. According to Ohmae, these are "natural economic zones which may or may not fall within the borders of individual nations." He states,

> Where prosperity exists, it is region-based. And when a region prospers, its good fortune spills over to adjacent territories inside and outside the political federation of which it is a part.
>   Region states are not—and need not be—the enemies of central government. Handled gently, by federation, these ports of entry to the global economy may well prove to be their very best friends.[26]

THE DIALECTIC OF "ISMS"

It is difficult to see beyond the nationalist economic paradigm. It is embedded within our theories, and our theories are embedded within the paradigm. It is the truth we know and the framework through which we understand the world.

The perspective of history and historical change gains us some purchase to stand outside and reexamine this perception of "reality." City-states were centers of commerce and wealth. City-states, however, lost the struggle for supremacy to nation-states and nationalistic mercantilism. The synthesis of these two was the nationalistic economic paradigm and its fusing of politics and economics in the political economy of statism. Through this statist lens, nations and national economies are one. Furthermore, in modified nationalism, states and local jurisdictions are also viewed as economies, always with jurisdictional and economic boundaries congruent.

Although the nationalist economic paradigm prevails in economic thought and political action, it has given rise to its own antithesis; that is, the pure "economism" of Jacobs and Ohmae, who correctly recognize that polities are not economies. Nevertheless, Jacobs's and Ohmae's views are essentially flawed both in overlooking the relevance of political jurisdictions and in denying significant roles for polities in economies.

In this book, we search for a new synthesis of the increasingly fatigued nationalist (statist) paradigm and the opposite extreme of economism. We try to frame a new political economy that recognizes (a) that polities and economies are not the same and (b) that political systems have an essential role in economies and vice versa. The hallmark of this new political economy is that it re-fuses politics and economics in a more realistic, functional way.

NOTES

1. Greer, S. (1962). *The emerging city: Myth and reality.* New York: MacMillan Press. p. 198.

2. Braudel, F. (1984). *Civilization and capitalism: 15th-18th century* (Vol. III). New York: Harper & Row. (Originally published in French as *Le Temps du Monde* [1979]. Paris: Librairie Armand Colin)

3. Braudel, F. (1984). *Civilization and capitalism: 15th-18th century* (Vol. III, p. 91). New York: Harper & Row. (Originally published in French as *Le Temps du Monde* [1979]. Paris: Librairie Armand Colin)

4. Braudel, F. (1984). *Civilization and capitalism: 15th-18th century* (Vol. III, p. 91). New York: Harper & Row. (Originally published in French as *Le Temps du Monde* [1979]. Paris: Librairie Armand Colin)

5. Braudel, F. (1984). *Civilization and capitalism: 15th-18th century* (Vol. III, pp. 311-312). New York: Harper & Row. (Originally published in French as *Le Temps du Monde* [1979]. Paris: Librairie Armand Colin)

6. Braudel, F. (1984). *Civilization and capitalism: 15th-18th century* (Vol. III, p. 322). New York: Harper & Row. (Originally published in French as *Le Temps du Monde* [1979]. Paris: Librairie Armand Colin)

7. Braudel, F. (1984). *Civilization and capitalism: 15th-18th century* (Vol. III, p. 194). New York: Harper & Row. (Originally published in French as *Le Temps du Monde* [1979]. Paris: Librairie Armand Colin)

8. Braudel, F. (1984). *Civilization and capitalism: 15th-18th century* (Vol. III, p. 287). New York: Harper & Row. (Originally published in French as *Le Temps du Monde* [1979]. Paris: Librairie Armand Colin)

9. Braudel, F. (1984). *Civilization and capitalism: 15th-18th century* (Vol. III, p. 288). New York: Harper & Row. (Originally published in French as *Le Temps du Monde* [1979]. Paris: Librairie Armand Colin)

10. Gray, A. (1931). *The development of economic doctrine.* New York: John Wiley.

11. Bender, T. (1983, Winter). The end of the city? *Democracy, 3*(1).

12. Bender, T. (1983, Winter). The end of the city? *Democracy, 3*(1).

13. Thurow, L. (1980). *Zero sum society* (p. 16). New York: Basic Books.

14. Jacobs, J. (1984). *Cities and the wealth of nations: Principles of economic life* (p. 180). New York: Random House. Braudel's term is "high voltage urban economy."

15. Jacobs, J. (1984). *Cities and the wealth of nations: Principles of economic life* (p. 3). New York: Random House.

16. Jacobs, J. (1984). *Cities and the wealth of nations: Principles of economic life* (pp. 6-7). New York: Random House.

17. Jacobs, J. (1984). *Cities and the wealth of nations: Principles of economic life* (p. 29). New York: Random House.

18. Jacobs, J. (1984). *Cities and the wealth of nations: Principles of economic life* (p. 31). New York: Random House.

19. Jacobs, J. (1984). *Cities and the wealth of nations: Principles of economic life* (p. 30). New York: Random House.

20. Jacobs, J. (1984). *Cities and the wealth of nations: Principles of economic life* (p. 31). New York: Random House.

21. Jacobs, J. (1984). *Cities and the wealth of nations: Principles of economic life* (p. 32). New York: Random House.

22. Ohmae, K. (1991). *The borderless world: Power and strategy in the interlinked economy* (p. 18). New York: Harper Perennial.

23. Ohmae, K. (1995). *The end of the nation state: The rise of regional economics* (p. viii). New York: Free Press.

24. Ohmae, K. (1995). *The end of the nation state: The rise of regional economics* (p. 142). New York: Free Press.

25. Ohmae, K. (1995). *The end of the nation state: The rise of regional economics* (p. 8). New York: Free Press.

26. Ohmae, K. (1995). *The end of the nation state: The rise of regional economics* (pp. 79-100). New York: Free Press.

27. Ohmae, K. (1995). *The end of the nation state: The rise of regional economics* (p. 100). New York: Free Press.

THE ECONOMIC REGION

Two themes emerge. First, economic paradigms are not timeless or unchanging. Nor are the frameworks of political economies. After politics gained ascendancy over economics in the historical triumph of nation-states over city-states, politics and economics were fused through the prevailing paradigm of the national economy identifying the "nation as the economy" or the "economy as the nation."

Second, political jurisdictions are not economies, just as economies are not political jurisdictions. Jane Jacobs has been a distinctive voice challenging the linkage of nation and economy in nationalist economics. Recently, Kenichi Ohmae's thesis of the withering of the nation-state and the rise of regional economies also takes on the prevailing paradigm of economics.

We are left with a thesis and antithesis: nationalist economies with their fusing of nation and economy and regional economies with no significant role for national governments. This presents an unfortunate and unproductive choice between statism and economism. To move forward, a synthesis is needed that constitutes a new political economy

embodying a new fusion of economics and politics in which governments, including national governments, play important and decisive
roles.

Before turning to this synthesis in Chapters 6 through 8, it is necessary to address the obvious question: If jurisdictions are not economies,
then what are the real economies? The central argument of this chapter
and Chapters 4 and 5 is that local economic regions are the basic economic units and the building blocks of the U.S. economy. The nationalist
economic paradigm cannot account for this reality.

WHAT IS A REGION?

The idea that the United States is segmented into distinctive regions,
and that these regions have important implications for public policy, is
certainly not new.

The earliest attempt to use data to delineate regions in the United
States appeared in two tables of the 1850 census report that presented
educational statistics for five geographical divisions: New England
states, middle states, southern states, southwestern states, and northwestern states.[1] The 1860 census included agricultural statistics for five
regions. It also contained one table showing population and density per
square mile for a set of geographic areas: New England states, middle
states (Maryland, Delaware, and Ohio), coast planing states (South
Carolina, Georgia, Florida, Alabama, Mississippi, and Louisiana), central slave states (Virginia, North Carolina, Tennessee, Kentucky, Mis-
souri, and Arkansas), northwestern states (Indiana, Illinois, Michigan,
Wisconsin, Iowa, and Minnesota), and, separately, Texas and California.

The concept of "vernacular regions" was introduced by Zelinsky in
1980.[2] The vernacular region is a "shared, spontaneous image of terri-
torial reality." Zelinsky identifies 14 vernacular regions that, he argues,
are important to understanding major issues in our society today.
Recently, Joel Garreau argued that North America is, in fact, nine
nations. Each of these, Garreau asserted, has its own capital and distinctive web of power and influence, with the "national" borders often not
following any existing political boundaries.[3] Public policy, according to
Garreau, is replete with failures because of our ignorance of the increasing power and independence of these regions.

The 1970s gave rise to rhetoric about the "new war between the states" and to new terms such as *Sun Belt, Frost Belt,* and *Rust Belt.* This discourse (or lack thereof) reflected real and perceived regional economic differences and rates of economic growth. Studies of regional economic performance have also documented consistently high degrees of variation deriving from differences in competitive factors.[4]

There have been periodic calls, some as early as the 1880s, for restructuring the geography of states to better reflect the social and economic realities of the national geography.[5] In the early 1960s, Rexford Tugwell proposed a massive restructuring of the federal system. Under this design, the United States would become a commonwealth composed of 12 or more regions.[6]

The emergence of the urban region has been recognized in proposals for restructuring a federal system that is "radically out of kilter with the demographic, economic and social realities of present day American life."[7] The creation of the areawide Municipality of Metropolitan Toronto in Canada, a federation of existing local governments, is an example of institutional changes mirroring the urban region. A more radical proposal in 1970 was to make metropolitan areas with populations of 1 million or more into states.[8]

Each of these proposals reflects the recognition that the functional area of social and economic organization greatly exceeds the span of political control of political jurisdictions. The latter proposal for consolidated metropolitan regions closely mirrors the concept of the comprehensive metropolitan region as a functional subnational economic area.

The geographical delineation of the United States that most closely corresponds to the idea of urban-centered economic regions is the Bureau of Economic Analysis's (BEA) classification of economic areas. The BEA defines the "economic area" as a functional, nodal unit composed of all economic components of the area. Each area contains one or more urban centers and the surrounding counties whose economic activity is focused on them; each area also contains a group of heterogeneous industries. The BEA definition combines place of work and place of residence for the economic area's labor force to minimize the effects of commuting across area boundaries. In addition, the economic areas have both export-based and residentially based industries.

The following salient features of the 172 economic areas defined by BEA should be noted:

1. The BEA delineation of economic areas exhausts the geography of the United States.
2. The BEA economic areas meet the chief criteria for metropolitan economic regions—that is, they are metropolitan centered with a surrounding hinterland polarized toward the dominant urban hub.
3. These functional economic areas do not correspond to any existing local government jurisdictions nor do they necessarily respect state boundaries (although the subunits are all counties) (Figure 3.1).
4. The BEA's geoeconomic articulation is bounded by the national jurisdiction of the United States.

The fourth feature of the system of BEA economic areas may represent a drawback, and perhaps a serious one, in examining metropolitan economic regions. There is no a priori reason why economic regions should conform to the political and economic boundaries of a nation. In the absence of national jurisdictions, it is almost certain that functional economic areas would spill across these economically artificial barriers. It is probably undeniable, however, that national boundaries and politically imposed restrictions effectively and functionally diminish and, in some cases, sever economic linkages within and among metropolitan economic regions—for example, the fragmentation of the Berlin economic region before the reunification of West and East Germany.

A second, but closely related, regional delineation comes from the Bureau of Labor Statistics (BLS), which defines labor market areas (LMAs). Each LMA consists of a central city and its contiguous areas that are economically integrated into that city. In general, workers can change jobs within the LMA without relocating. LMAs are defined in terms of counties; usually, major LMAs and metropolitan statistical areas (MSAs) have the same boundaries. Small LMAs are composed of a county, or a group of counties, with a central community of 5,000 or more inhabitants and within which workers can commute. The land area of each state is divided into its LMAs; as a result, they are contiguous and usually do not cross state boundaries. Employment and unemployment data collected by the local Department of Labor offices are available by LMAs.[9]

Data availability is a difficult and often defining issue in attempting to delineate economic regions. For the most part, data are available only for those geographic areas that have political boundaries. The majority of subnational or substate areas have been defined primarily by political jurisdiction in terms largely unrelated to economic patterns.

Nevertheless, an unmistakable pattern emerges in that both the BEA's economic areas and the BLS's labor market areas are urban centered. This is a reflection of the fact that population and economic activity in the United States are centralized in metropolitan areas, suggesting a strong relationship between economic growth and urbanization.

By 1990, more than three fourths of the nation's population lived in metropolitan areas compared to 40% at the turn of the century. Whereas the population of the United States has tripled during the course of this century, the population of metropolitan areas has increased sixfold. In contrast, the population of nonmetropolitan areas has grown more slowly than the total population, increasing by only 23% since 1900 (Figure 3.2).[10]

Even these measures, however, fail to capture the true extent of the urbanization of the United States. In 1986, 10% of the population lived in nonmetropolitan counties that claimed "urbanized" populations of 20,000 or more and that were adjacent to metropolitan areas.[11] Thus, 87% of the nation's population lived in metropolitan areas or adjacent urbanized counties. An additional 10% of the population was in urbanized counties that were not adjacent to metropolitan areas. In 1986, therefore, slightly more than 97% of the population of the United States was urbanized (Figure 3.3).

Economic activity is even more concentrated in metropolitan areas than is population. Between 1980 and 1986, metropolitan areas experienced higher rates of population and employment growth, but the rate of job growth was almost three times greater than the increase in population. During the same period, approximately 94% of the total increase in employment in the United States occurred in metropolitan areas.[12] Urbanized adjacent areas accounted for an additional 3% and urbanized nonadjacent areas for an additional 2%. Less than 1% of net job growth in the United States between 1980 and 1986 occurred in totally rural counties.

This ongoing centralization of economic activity into urbanized areas suggests that economic growth and urbanization are closely tied. Metropolitan areas are responsible for almost all net new jobs created in the

(text continues on page 28)

Figure 3.1. BEA Map

U.S. Department of Commerce
Bureau of Economic Analysis

Codes and Names for BEA Economic Areas

Code	Name	Code	Name
001	Bangor, ME	044	Knoxville, TN
002	Portland, ME	045	Johnson City-Kingsport-Bristol, TN-VA
003	Boston-Worcester-Lawrence-Lowell-Brockton, MA-NH-RI-VT	046	Hickory-Morganton, NC-TN
004	Burlington, VT-NY	047	Lexington, KY-TN-VA-WV
005	Albany-Schenectady-Troy, NY	048	Charleston, WV-KY-OH
006	Syracuse, NY-PA	049	Cincinnati-Hamilton, OH-KY-IN
007	Rochester, NY-PA	050	Dayton-Springfield, OH
008	Buffalo-Niagara Falls, NY-PA	051	Columbus, OH
009	State College, PA	052	Wheeling, WV-OH
010	New York-North New Jersey-Long Island, NY-NJ-CT-PA-MA-VT	053	Pittsburgh, PA-WV
011	Harrisburg-Lebanon-Carlisle, PA	054	Erie, PA
012	Philadelphia-Wilmington-Atlantic City, PA-NJ-DE-MD	055	Cleveland-Akron, OH-PA
013	Washington-Baltimore, DC-MD-VA-WV-PA	056	Toledo, OH
014	Salisbury, MD-DE-VA	057	Detroit-Ann Arbor-Flint, MI
015	Richmond-Petersburg, VA	058	Northern Michigan, MI
016	Staunton, VA-WV	059	Green Bay, WI-MI
017	Roanoke, VA-NC-WV	060	Appleton-Oshkosh-Neenah, WI
018	Greensboro-Winston-Salem-High Point, NC-VA	061	Traverse City, MI
019	Raleigh-Durham-Chapel Hill, NC	062	Grand Rapids-Muskegon-Holland, MI
020	Norfolk-Virginia Beach-Newport News, VA-NC	063	Milwaukee-Racine, WI
021	Greenville, NC	064	Chicago-Gary-Kenosha, IL-IN-WI
022	Fayetteville, NC	065	Elkhart-Goshen, IN-MI
023	Charlotte-Gastonia-Rock Hill, NC-SC	066	Fort Wayne, IN
024	Columbia, SC	067	Indianapolis, IN-IL
025	Wilmington, NC-SC	068	Champaign-Urbana, IL
026	Charleston-North Charleston, SC	069	Evansville-Henderson, IN-KY-IL
027	Augusta-Aiken, GA-SC	070	Louisville, KY-IN
028	Savannah, GA-SC	071	Nashville, TN-KY
029	Jacksonville, FL-GA	072	Paducah, KY-IL
030	Orlando, FL	073	Memphis, TN-AR-MS-KY
031	Miami-Fort Lauderdale, FL	074	Huntsville, AL-TN
032	Fort Meyers-Cape Coral, FL	075	Tupelo, MS-AL-TN
033	Sarasota-Bradenton, FL	076	Greenville, MS
034	Tampa-St. Petersburg-Clearwater, FL	077	Jackson, MS-AL-LA
035	Tallahassee, FL-GA	078	Birmingham, AL
036	Dothan, AL-FL-GA	079	Montgomery, AL
037	Albany, GA	080	Mobile, AL
038	Macon, GA	081	Pensacola, FL
039	Columbus, GA-AL	082	Biloxi-Gulfport-Pascagoula, MS
040	Atlanta, GA-AL-NC	083	New Orleans, LA-MS
041	Greenville-Spartanburg-Anderson, SC-NC	084	Baton Rouge, LA-MS
042	Asheville, NC	085	Lafayette, LA
043	Chattanooga, TN-GA	086	Lake Charles, LA
		087	Beaumont-Port Arthur, TX
		088	Shreveport-Bossier City, LA-AR
		089	Monroe, LA
		090	Little Rock-North Little Rock, AR
		091	Fort Smith, AR-OK
		092	Fayetteville-Springdale-Rogers, AR-MO-OK

Codes and Names for BEA Economic Areas

Code	Name	Code	Name
093	Joplin, MO-KS-OK	134	San Antonio, TX
094	Springfield, MO	135	Odessa-Midland, TX
095	Jonesboro, AR-MO	136	Hobbs, NM-TX
096	St. Louis, MO-IL	137	Lubbock, TX
097	Springfield, IL-MO	138	Amarillo, TX-NM
098	Columbia, MO	139	Santa Fe, NM
099	Kansas City, MO-KS	140	Pueblo, CO-NM
100	Des Moines, IA-IL-MO	141	Denver-Boulder-Greeley, CO-KS-NE
101	Peoria-Pekin, IL	142	Scottsbluff, NE-WY
102	Davenport-Moline-Rock Island, IA-IL	143	Casper, WY-ID-UT
103	Cedar Rapids, IA	144	Billings, MT-WY
104	Madison, WI-IL-IA	145	Great Falls, MT
105	La Crosse, WI-MN	146	Missoula, MT
106	Rochester, MN-IA-WI	147	Spokane, WA-ID
107	Minneapolis-St. Paul, MN-WI-IA	148	Idaho Falls, ID-WY
108	Wausau, WI	149	Twin Falls, ID-WY
109	Duluth-Superior, MN-WI	150	Boise, ID-OR
110	Grand Forks, ND-MN	151	Reno, NV-CA
111	Minot, ND	152	Salt Lake City-Ogden, UT-ID
112	Bismark, ND-MT-SD	153	Las Vegas, NV-AZ-UT
113	Fargo-Moorhead, ND-MN	154	Flagstaff, AZ-UT
114	Aberdeen, SD	155	Farmington, NM-CO
115	Rapid City, SD-MT-NE-ND	156	Albuquerque, NM-AZ
116	Sioux Falls, SD-IA-MN-NE	157	El Paso, TX-NM
117	Sioux City, IA-NE-SD	158	Phoenix-Mesa, AZ-NM
118	Omaha, NE-IA-MO	159	Tucson, AZ
119	Lincoln, NE	160	Los Angeles-Riverside-Orange County, CA-AZ
120	Grand Island, NE	161	San Diego, CA
121	North Platte, NE-CO	162	Fresno, CA
122	Wichita, KS-OK	163	San Francisco-Oakland-San Jose, CA
123	Topeka, KS	164	Sacramento-Yolo, CA
124	Tulsa, OK-KS	165	Redding, CA-OR
125	Oklahoma City, OK	166	Eugene-Springfield, OR-CA
126	Western Oklahoma, OK	167	Portland-Salem, OR-WA
127	Dallas-Fort Worth, TX-AR-OK	168	Pendleton, OR-WA
128	Abilene, TX	169	Richland-Kennewick-Pasco, WA
129	San Angelo, TX	170	Seattle-Tacoma-Bremerton, WA
130	Austin-San Marcos, TX	171	Anchorage, AK
131	Houston-Galveston-Brazoria, TX	172	Honolulu, HI
132	Corpus Christi, TX		
133	McAllen-Edinburg-Mission, TX		

Figure 3.1. BEA Map Legend

Note: The 172 BEA economic areas are defined as of February 1995. Codes are assigned, beginning with 001 in northern Maine, continuing south to Florida, then north to the Great Lakes, and continuing in a serpentine pattern to the West Coast. Except for the western Oklahoma economic area (126), the northern Michigan economic area (058), and the 17 economic areas corresponding to CMSAs, each economic area is named for the metropolitan area or city that is the node of its largest CEA and that is usually, but not always, the largest metropolitan area or city in the economic area. The name of each economic area includes each state that contains counties in that economic area.

Population
(in thousands)

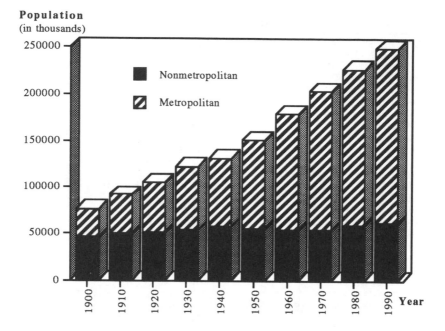

Figure 3.2. Metropolitan and Nonmetropolitan Distribution of Population in the United States, 1900-1990. Numbers prior to 1960 exclude Alaska and Hawaii.
SOURCE: U.S. Department of Commerce, Bureau of the Census, Washington, D.C.

economy. They are home to more than three fourths of the nation's population. They are growing faster than the nation as a whole. Also, all this is occurring on only 16% of the United States's geographical territory.

Demographic data alone, however, are insufficient to show that the urban place is a causal force in the process of economic growth. To explore this thesis further, we must look beyond demographics to economic theory and research.

ORIGINS OF ECONOMIC GROWTH: THE PRODUCTIVE METROPOLIS

Economists, throughout the development of the discipline, have largely ignored the relationship between urban places and economic growth

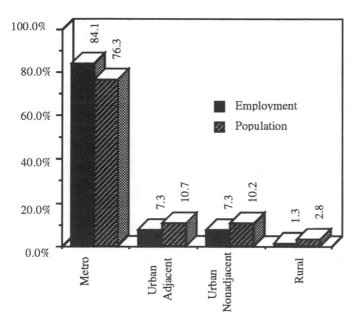

Figure 3.3. Urbanization of Employment and Population in the United States, 1986 (Percentage distribution by level of urbanization)

and productivity. John R. Harris argues that the city's dominant image in economics has been as the invisible city.[13] At best, he argues, urbanization traditionally was viewed as a concomitant of structural change rather than a causal force in industrialization. Systematic consideration of the functions of the urban place in the process of industrialization did not occur until the 1950s, when seminal articles by Eric Lampard, Bert Hoselitz, and Wilbur Thompson focused on the relationship between urbanization and economic growth and development.[14] For the first time, the urban place became visible in economics.

To a significant degree, this literature is focused on the city as the functional economy rather than on the metropolis that has emerged as the functional economy during the past four decades. Nevertheless, Lampard, Hoselitz, Thompson, and others offer three primary economic images of the metropolis that are quite relevant to our discussion. The first is that of the productive metropolis in which proximity, size,

and resource concentrations provide economies of scale and competitive cost advantages. The second economic image is that of the generative metropolis, a seedbed and incubator for entrepreneurship and
innovation. The third is the image of the metropolis of control—the
center of decision making, information, and communications.

The productive metropolis is the realization of agglomerative economies deriving from a variety of factors. Among these are the minimiza-
tion of the costs of distance, including transportation, communications,
and time; common factors of production such as pools of skilled and
unskilled labor and suppliers of common inputs; complementary stages
of production; coordination of specialized activities permitted by proximity; and availability of social overhead capital.

In its generative mode, the metropolis nurtures innovation and
entrepreneurship, which in turn spur cumulative growth. Jane Jacobs's
thesis of cities as sources of economic energy in a national economy, first
introduced in *The Death and Life of Great American Cities* and later the
centerpiece of *Cities and the Wealth of Nations*, is predicated on the role
of cities in fostering innovations that drive economic growth. The causal
relationship between the metropolis and innovation, however, remains
elusive. Jacobs argues that cities serve as incubators because of the
interaction they facilitate among inventors and entrepreneurs. Others
assert the importance of technical suppliers that reduce the capital
requirements for innovation.

The difficulties in attempting to quantify and test these assertions are
severe and efforts to do so have been inconclusive.[15] Nevertheless,
existing tests, which focus primarily on the incubation of small firms
within central cities and their diffusion across the wider metropolitan
area, do not refute the important role of the metropolis in promoting
entrepreneurship and innovation.

The control function of the metropolis derives from the centralization
in the urban area of specialized information, communications, interaction, and advanced transportation. The spatial proximity of these
functions creates specialized environments for administrative headquarters (public and private), financial intermediaries, data processing,
and specialized business services. Of particular importance is the concentration in the metropolis of headquarters of multilocational business
organizations that control the majority of jobs throughout the U.S.
economic system.[16]

The concept of the metropolitan region as a primary source of developmental change, productivity, and growth links the theories of urban growth centers and regional growth in a national urban system. This concept is consistent with contemporary views of urban ecology that

> Economic growth takes place in a matrix of urban regions through which the space economy is organized. The crux of the link between regional growth and modern growth center concepts is that it is cities within the urban system, linked by filtering mechanisms—not the heartland-hinterland lever in the regional system, linked by export-base multipliers, that today organize the economy spatially. The cities are centers of activity and of innovation, focal points of the transport and communications networks, locations of superior accessibility at which firms can most easily reap scale economies and at which industrial complexes can obtain the economies of the localization and urbanization. They encourage labor specialization, areas specializing in productive activities, and efficiency in the provision of services. Agricultural enterprise is more efficient in the vicinity of cities. The more prosperous commercialized agricultures encircle the major cities, whereas the peripheries of the great urban regions are characterized by backward lower-income economic systems.[17]

VARIATIONS ACROSS URBAN REGIONAL ECONOMIES

Niles Hansen stated that the United States abandoned a national perspective on regional development policy issues in the 1980s in part because of the neoclassical assumption that free market forces will narrow economic disparities among regions. Echoing Jacobs's challenge to the national orientation of federal policy, Hansen argued that

> There is no such thing as the American economy. The United States is rather a collection of heterogeneous regions with differing problems and opportunities. This has important implications for economic policy.
> National policies and programs that implicitly treat the United States as a homogeneous area or that are oriented toward some "average" situation thereby fail to take into account the fact that the nation is a collection of heterogeneous regions.[18]

The degree of variation in the economic performance of regions in the United States is inconsistent with the assumption of a single, somewhat homogeneous national economy. Evidence of the extent of this variation suggests that regional economies may be the basic economic building blocks.

The degree of variation in the performance of metropolitan economies is easily demonstrated. Figures 3.4 through 3.6 examine median household incomes and unemployment and poverty rates for the 50 largest metropolitan areas in 1990. These figures also identify the comparable national levels, as well as the metropolitan averages, clearly demonstrating the extent of variation regarding the average national measures that underlie federal policy and program decisions.

In 1989, median household income was $30,056 in the nation and $32,056 in the metropolitan sector. The variation of large metropolitan areas regarding these averages is illustrated in Figure 3.4. Median household incomes in these 50 largest metropolitan areas ranged from a high of $46,848 to a low of $24,442—a difference of $22,406.

Eight metropolitan areas fell below the national and metropolitan averages and 42 were above. Of these higher-income metropolitan areas, 18 exceeded the national average by more than $5,000, 10 by $10,000 or more, and 5 by $15,000 or more. Therefore, the range of variation in metropolitan median household incomes is significant, and the national averages mask the extent of this variation. Moreover, the diversity in metropolitan income averages is not related to metropolitan area size, at least within the 50 largest areas examined in Figure 3.4.

The unemployment rate is a primary measure of the performance of the national economy and its constituent metropolitan economies. In 1994, the national rate of unemployment averaged 6.1% (Figure 3.5). The average for the 50 metropolitan areas was 5.7%. Again, the range in unemployment rates across these metropolitan areas was wide, from a high of 9.4% to a low of 3.3%. The majority (33) experienced unemployment rates below the national and metropolitan averages, but the remaining 17 had rates in excess of the national mean. As with the income averages, there appears to be no consistent relationship between urban size and unemployment performance.

The capacity of the nation and its urban and rural economies to provide jobs and meaningful incomes is a critical measure of economic performance. The incidence of poverty in urban and rural places is one indicator of lagging economic performance or, at least, of the inability of the local economy to integrate some components of the population.

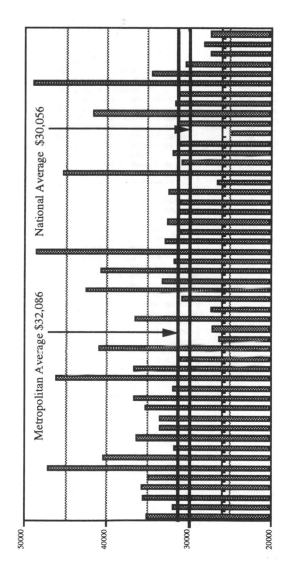

Figure 3.4. Median Household Income, 1989: Metropolitan Areas

33

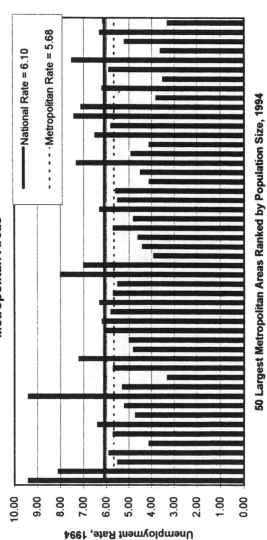

Figure 3.5. Unemployment Rates, 1994: Metropolitan Areas

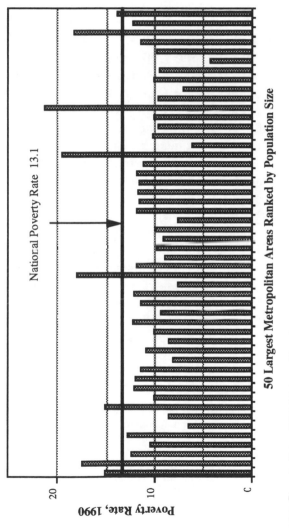

Figure 3.6. Poverty Rates, 1990: Metropolitan Areas

In 1990, 13.1% of the population lived in poverty. In metropolitan areas, the poverty rate was 12.1%.

The incidence of poverty in nonmetropolitan areas, therefore, continues to exceed that experienced in metropolitan areas. Nonetheless, the majority of those living in poverty in the United States reside in metropolitan areas.

The incidence of poverty varies greatly among metropolitan areas (Figure 3.6). Metropolitan poverty rates in 1990 varied from a high of 21.2% to a low of 4.2%. Among the 50 largest metropolitan areas, 7 had poverty rates in excess of 15%, 8 experienced poverty at rates above the national average, and the rate of poverty in 11 exceeded the average rate for the metropolitan sector. Conversely, 41% had poverty rates below the national average and 39 were below the metropolitan average.

Almost all measures of economic performance reveal similar patterns of variation among metropolitan regional economies. Variations in local business cycles offer perhaps the most salient example of variation among local economies (see Chapter 5). The breadth and depth of these patterns of metropolitan performance indicate that economic thinking must be sensitive to the diversity of urban places and their unique circumstances and needs.

From within the nationalist economic paradigm, it may be very difficult to accept regional economies as the real economies and the basic building blocks of the national and global economic systems. The variability of the performance of these metropolitan-centered economic regions, however, makes it difficult to sustain the proposition of a single, somewhat homogeneous national economy.

NOTES

1. See U.S. Department of Commerce, Bureau of the Census. (1981, June 5). *Census regions and census divisions* (Memorandum for the record). Washington, DC: Author.

2. Zelinsky, W. (1980). North America's vernacular regions. *Annals of the Association of American Geographers, 70*(1), 1.

3. Garreau, J. (1981). *Nine nations of North America* (p. 9). Boston: Houghton Mifflin.

4. Hansen, N. (1988). Economic development and regional heterogeneity: A reconsideration of regional policy for the United States. *Economic Development Quarterly, 2*(2), 107-118.

5. See for example: Patton, S. N. (1888, July). The decay of state and local governments. *Annals of the American Academy of Political Science, 1*(1), 26-42; Burgess, J. W. (1886,

March). The American commonwealth: Changes in its relation to the nation. *Political Science Quarterly, 1,* 9-35.

6. Tugwell, R. G. (1972). *Reforming the Constitution: An imperative for modern America.* Clio Press: Oxford.

7. Canty, D. (1972, March/April). Metropolity. *City.* Canty suggests (p. 39) that some will find proposals such as these, including his own for an elective Metropolitan Development Agency, "akin to swimming not just upstream, but up the face of a raging waterfall."

8. Burton, R. P. (1970). *The metropolitan state.* Washington, DC: The Urban Institute. (Reprinted from testimony before the Subcommittee on Urban Affairs of the Joint Economic Committee)

9. Another regional classification is Rand McNally's basic trading areas (BTA). A BTA is comprised of a basic trading center, or city, and the areas surrounding it that follow county boundaries. A BTA includes those areas with populations consisting of residents who buy most of their shopping goods from the center or its suburbs. These shopping goods include general merchandise that a shopper will compare to others in quality, styles, and prices before buying. Extending the notion of this type of area further, Rand McNally also has similar data for major trading areas (MTAs), which consist of two or more BTAs. Each of the MTAs is named after one or more cities that are its major trading centers and that serve as the area's primary center of wholesaling, distribution, and banking. Available data for both BTAs and MTAs include population, income, and retail sales.

10. Clearly, this increasing concentration of population in metropolitan areas is not solely attributable to greater density in the cities that existed at the turn of the century. Indeed, new metropolitan areas have been defined as they passed critical population thresholds. Furthermore, additional counties were added to existing MSAs as they became increasingly integrated with adjacent urban areas. In 1950, there were 169 metropolitan areas occupying 5.9% of the nation's land area. By 1988, the number had climbed to 282 metropolitan areas encompassing 16.2% of the nation's geography. Urbanization of the United States can thus be measured not only by population in metropolitan areas but also by the urbanization of the nation's geography.

11. The U.S. Department of Agriculture has developed a rural-urban continuum that classifies all counties by nine levels of urbanization or ruralization based on the 1980 census. For present purposes, these are collapsed into four categories of counties: those in metropolitan areas, those with urbanized populations of 20,000 or more adjacent to metropolitan areas, those with urbanized populations of 20,000 or more not adjacent to metropolitan areas, and those that are totally rural.

12. During the period from 1980 to 1986, the net increase in metropolitan population was greater than the net increase in total population, indicating that net immigration to metropolitan areas was occurring.

13. Harris, J. R. (1984). Economics: Invisible, productive, and problem cities. In L. Rodwin & R. M. Hollister (Eds.), *Cities of the mind: Images and themes of the city in the social sciences.* New York: Plenum.

14. Lampard, E. E. The evolving system of cities in the United States: Urbanization and economic development. In H. S. Perloff & L. Wingo, Jr. (Eds.), *Issues in urban economics.* Baltimore, MD: Johns Hopkins University Press; Lampard, E. E. (1955, January). The history of cities in the economically advanced areas. *Economic Development and Cultural Change;* Hoselitz, B. (1953, June). The role of cities in the economic growth of underdevel-

oped countries. *Journal of Political Economy*; Thompson, W. Internal and external factors in the development of urban economies. In H. S. Perloff & L. Wingo, Jr. (Eds.), *Issues in urban economics*. Baltimore, MD: Johns Hopkins University Press.

15. See for example: Leone, R. A., & Struyk, R. (1976, October). The incubator hypothesis: Evidence from five SMSAs. *Urban Studies, 13*(3), 48-57; Struyk, R., & James, F. (1976, October). *Intrametropolitan industrial location* (Urban and Regional Study No. 3). New York: National Bureau of Economic Research.

16. This argument is developed in detail in Pred, A. (1977). *City systems in advanced economies: Past growth, present processes and future development options* (Chap. 3). New York: John Wiley.

17. Berry, B., & Kasarda, K. (1977). *Contemporary urban ecology* (p. 279). New York: Macmillan.

18. Hansen, N. (1988). Economic development and regional heterogeneity: A reconsideration of regional policy for the United States. *Economic Development Quarterly, 2*(2), 112-118.

THE INTERNAL INTERDEPENDENCE OF REGIONS

Chapter 3 marshaled evidence that the nationalist economic view does not adequately explain economic reality in the United States. There are strong indications that distinct, metropolitan-centered regional economies exist in the United States and, therefore, that national average indicators of economic performance mask important differences in the performance of these regional economies.

There is also evidence of an internal economic coherence and interdependence within each of these local economic regions. This chapter and Chapter 5 push this line of argument forward.

The essential coherence of the local economic area is too often obscured in public and scholarly discussion by a focus on the relationship (or lack of relationship) between central cities and their suburbs. Locked in a debate that uses governmental jurisdictions as the units of analysis, protagonists and observers alike repeat the error of the nationalist paradigm at the local level—they equate the economic with the govern-mental. Through that lens, the economy has no distinct boundaries other than the governmental jurisdiction's borders.

This fundamental equating of economic and governmental units of analysis reproduces itself in data collection and reporting. Most eco-nomic data are arranged by governmental units—municipalities and counties. (The analysis in this chapter and Chapter 5 uses such data.)

What we have called the "nationalist" paradigm is thus, more gen-erically, a "statist" paradigm. It defines economies in terms of "the state" at every level—federal (i.e., national), state (e.g., Alabama and Arizona), and local (municipalities and counties). States and local gov-ernments mimic the federal claim to "an economy," although their claims are, of course, to smaller territories. Their claims also are consid-erably weaker because they have weaker economic tools with which to enforce them. Nevertheless, these claims and the paradigm that guides them are captivating enough to shape the way in which issues are developed, discussed, and addressed.

The issue in question here is often formulated in public discussion in terms of whether suburbs can prosper and succeed regardless of the fate and fortunes of their central city. Framed in this way, the question is whether the central city economy causes the suburban one and whether the suburbs are dependent on the central city. The question assumes two independent entities.

The thesis of this book is quite different. We argue that there is an underlying local economic region of which the city and suburban governmental jurisdictions each comprise only parts. The jurisdictions are not separate economies; they are governmental overlays on the broader economic entity. The coherence of that economic region pro-duces city and suburban economic performances that are interdepen-dent. The specifics of that mutual interdependence will be different in each metropolitan area. The key is that city and suburb are parts of a larger economic whole.

In a 1993 report,[1] we used the 1990 median household income data to test the argument that the economic futures of cities and suburbs are intertwined. If cities and suburbs are economically independent, rather than interdependent, suburban and central-city incomes should not be related nor move together. If there is some degree of interdependence based in the shared local economic region, however, changes in subur-

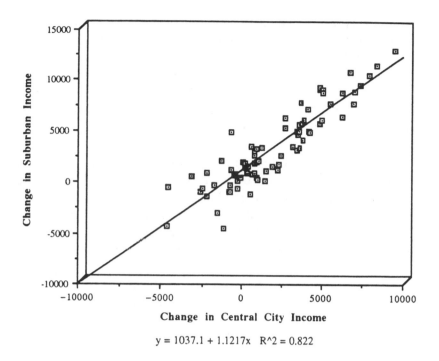

y = 1037.1 + 1.1217x   R^2 = 0.822

Figure 4.1. Change in Median Household Income, 1979-1989: Central Cities and Suburbs. Reprinted with permission from Barnes, W. R., & Ledebur, L. C. (1993). *All in it together, cities, suburbs and local economic regions.* Washington, DC: National League of Cities.

ban incomes would be related to changes in central-city incomes and vice versa.

The crucial test comes in examining the relationship of the change in median household income in cities and suburbs over time. If there is no relationship between suburban and central-city incomes, a plot of these relationships would appear randomly scattered. If, on the other hand, there is a strong relationship, the plot of these relationships, the scattergram, would cluster in a clear and discernible pattern.

This relationship is presented in Figure 4.1. The pattern is sharp and distinct. The relationship is very strong. The percentage of the variation explained is 82%. This is clear evidence of a strong and consistent relationship between changes in central-city and suburban incomes. The interpretation of this relationship is as follows. For every $1 increase in central-city income, suburban income increases by $1.12. Conversely, for every increase of $1.12 in suburban income, central-city income increases by $1.

The relationship does not imply causation—that is, that change in one causes the change in the other. Rather, the relationship is mutual, interactive, and interdependent. This evidence strongly suggests that the economic fates and fortunes of cities and suburbs are inextricably intertwined.

The data also suggest that where suburban incomes are increasing, central-city incomes also are on the upswing. Conversely, where central-city incomes are decreasing, suburban incomes are in decline. Table 4.1 presents data on these patterns of change for the 25 metropolitan areas in which suburban incomes increased most rapidly between 1980 and 1990. Table 4.2 focuses on the 18 metropolitan areas in which suburban incomes declined over this period.

In each of the 25 metropolitan areas with the most rapidly growing suburbs, measured by absolute and percentage gains in median household income, central-city incomes also increased during the 1979 to 1989 period. In other words, in this high-growth set, no suburbs experienced income growth without corresponding growth in their central cities. This also means that no central city in this high-growth sample experienced income growth in the absence of suburban growth. In high-growth areas, suburbs and central cities grew together. In all but 1 (San Diego) of the 25 metropolitan areas, the absolute gains in suburbs exceeded those in their central cities. In 10 of these metropolitan areas, however, the rate of central-city income growth exceeded that of the suburbs.

Over the 10-year period, suburbs in 18 of the 78 largest metropolitan areas experienced declines in real median household incomes (Table 4.2). In all but 4 of these, central-city incomes also declined. In only 1 of these 4 did the decline in suburban income exceed 1% (Salt Lake City: −4.7%). The corresponding increases in central-city incomes were also relatively small, ranging between 1% and 4%. Tables 4.1 and 4.2 simply confirm the findings of Figure 4.1. Suburban and central-city median household incomes tend to move together.

Of the remaining 35 metropolitan areas (those not presented in Tables 4.1 and 4.2), all experienced suburban income growth. Central-city median household income grew in 25 and declined in 10 areas.

TABLE 4.1   Metropolitan Areas With Rapidly Growing Suburbs:
Change in Median Household Income, 1979-1989

Metropolitan Area	Suburbs		Central City	
	Absolute ($)	%	Absolute ($)	%
Bridgeport-Stamford-Norwalk-Danbury, CT	12,519	26.1	9,196	29.8
Oxnard-Ventura, CA PMSA	10,995	28.2	8,095	26.4
New York, NY PMSA	10,395	26.3	6,464	27.7
San Jose, CA PMSA	10,096	25.1	7,620	19.7
Boston-Lawrence-Salem-Lowell-Brockton, MA	9,083	26.3	7,087	32.2
Middlesex-Somerset-Hunterdon, NJ PMSA	8,906	21.4	4,639	19.6
Newark, NJ PMSA	8,669	22.9	4,592	26.9
Hartford-New Britain-Middletown-Bristol, CT	8,650	22.8	4,790	19.4
Bergen-Passaic, NJ PMSA	8,433	21.7	6,729	33.3
Anaheim-Santa Ana, CA PMSA	8,372	21.0	6,729	14.8
New Haven-Waterbury-Meriden, CT NECMA	8,286	23.2	6,010	25.4
Washington, DC-MD-VA MSA	7,315	17.1	3,396	12.4
San Francisco, CA PMSA	7,237	18.2	6,663	24.9
Worcester-Fitchburg-Leominster, MA NECMA	7,234	22.7	5,306	21.7
Oakland, CA PMSA	6,687	18.2	3,862	16.6
San Diego, CA PMSA	5,914	19.7	6,021	21.8
Philadelphia, PA-NJ PMSA	5,880	16.7	2,401	10.8
Raleigh-Durham, NC MSA	5,672	18.4	4,756	18.5
Honolulu, HI MSA	5,593	14.7	3,644	10.9
Los Angeles-Long Beach, CA PMSA	5,351	16.4	4,586	17.3
Providence-Pawtucket-Woodstock, MA NECMA	5,301	17.5	3,472	16.9
Riverside-San Bernadino, CA PMSA	5,238	18.3	3,293	12.0
Baltimore, MD MSA	4,838	12.9	2,445	11.3
West Palm Beach-Boca Raton-Delray Beach, FL MSA	4,586	16.4	3,950	14.0
Albany-Schenectady-Troy, NY MSA	4,562	14.7	3,297	15.6

In the overall sample of 78 metropolitan areas, suburban incomes grew in 60 and declined in 18. Central-city incomes grew in 54 and declined in 24. The direction of change in suburban and central-city

TABLE 4.2 Metropolitan Areas With Suburbs Experiencing Income Declines: Change in Median Household Income, 1979-1989

Metropolitan Area	Suburbs		Central City	
	Absolute ($)	%	Absolute ($)	%
Fresno, CA MSA	(9)	0.0	600	2.5
Cincinnati, OH-KY-IN PMSA	(24)	−0.1	(363)	−1.7
Portland, OR PMSA	(204)	−0.6	669	2.7
Las Vegas, NV MSA	(281)	−0.9	1,139	3.9
Kansas City, MO-KS MSA	(314)	−0.9	(554)	−2.1
Buffalo, NY PMSA	(743)	−2.2	(1,063)	−5.4
Tucson, AZ MSA	(755)	−2.3	(2,001)	−8.4
Detroit, MI PMSA	(969)	−2.4	(4,830)	−20.5
Tulsa, OK MSA	(1,005)	−3.4	(2,737)	−9.6
Louisville, KY-IN MSA	(1,050)	−3.3	(552)	−2.7
Oklahoma City, OK MSA	(1,327)	−4.5	(1,079)	−4.0
Denver, CO PMSA	(1,347)	−3.5	(1,038)	−4.0
Cleveland, OH PMSA	(1,380)	−3.7	(2,878)	−13.9
Salt Lake City-Ogden, UT MSA	(1,628)	−4.7	224	1.0
Akron, OH PMSA	(1,765)	−5.0	(2,510)	−10.1
Pittsburgh, PA PMSA	(3,418)	−10.8	(1,859)	−8.2
Houston, TX PMSA	(4,667)	−11.0	(4,886)	−15.7
New Orleans, LA MSA	(4,913)	−14.7	(1,442)	−7.2

household incomes was the same in all but 14 of these metropolitan areas (18%). The bottom line is that central-city and suburban incomes generally go up and down together.

Of course, disparities in city and suburban income levels exist. The point, as Figure 4.1 shows, is that those disparate levels are yoked to each other; they go up or down together at varying rates. Although this documentation of the significance of the positive relationship between changes in suburban and central-city household incomes is important, it should not be surprising. Cities and their suburbs are not two distinct economies. They are parts of a single regional economy, highly interdependent and with their fortunes intertwined.

Other researchers have uncovered similar findings concerning the interdependence of cities and suburbs. Richard Voith of the Federal

Reserve Bank of Philadelphia researched the question of whether cities and suburbs could be classified as complements or substitutes. He concludes that "decline in central cities is associated with slow-growing suburbs," noting that this fact often goes unrecognized by suburban dwellers because "the suburb is performing so much better than its declining central city counterpart."[2]

H. V. Savitch and colleagues examined the relationships among income, jobs, and population characteristics for cities and their suburbs. These researchers found a strong association between income disparities and poverty, which led them to conclude that the "blight of the inner city casts a long shadow. Companies will not grow in, or move to, a declining environment."[3]

For the northeast and north central regions, Charles Adams found that

> strong cities appear to positively reinforce naturally evolving patterns of suburbanization. . . . Weaker central cities appear to exert a negative influence on metropolitan suburbanization, with a higher proportion of those migrating out of the central city moving to locations outside the SMSA.[4]

Hill, Wolman, and Ford reviewed this research. After a critique of the use of correlation analysis, they reached similar conclusions using more sophisticated statistical techniques. They identified and responded to the following three questions, which they believed to be at the heart of the issue of interdependence of metropolitan regions:

> First, are cities and suburbs interdependent? The answer is yes, because they are part of the same labor and land markets. Second, can there be healthy individual suburbs and weak central cities? Yes, it is a common— or even typical—occurrence. . . . Third, can there be healthy suburbs in the aggregate and extremely poor central cities? Once again, yes, but it is a rarer occurrence.[5]

Hill et al. conclude their analysis as follows:

> Distinctions between central cities and suburbs have little economic meaning as separable units of private production (with the prominent exception of the role of local taxes in business location decisions and production costs), income generation, or wealth creation. For these ac-

tivities, the relevant economic unit is the entire metropolitan area, because that is the geography that encompasses the functional labor and land markets.[6]

Others have combined the findings of these and related studies with their practical experiences in addressing these issues. For instance, Peter Dreier asserts that

> The reputation and viability of the entire metropolitan area is shaped by the impressions of the central city. Allowing the central city to decay affects the entire metropolitan area. Businesses are reluctant to move to, invest in, or remain in a metropolitan area where the central city has a high crime rate, an inferior school system, inadequate services, and the potential for civil disorder.[7]

Anthony Downs of The Brookings Institution notes that "no jurisdiction is an island. Every suburb is linked to its central city and to other suburbs."[8] We find this phrase especially appropriate because it alludes to the idea of a city-suburb interdependence while hinting at the relationship between polities and economies.

Henry Cisneros, former secretary of the U.S. Department of Housing and Urban Development, stated that

> Central cities and their suburbs are clearly "joined at the hip" in the structure and functioning of interrelated economic activities, stretching from the older downtown central business districts to the new "edge city" suburban office and retail centers and from inner city manufacturing plants to suburban industrial parts.[9]

In building the case that economic regions are the basic economic units and the building blocks of the greater economic systems, we first demonstrated the existence of significant diversity in the performance of regions rather than the more uniform performance implied by the nationalist paradigm of the economy (Chapter 3). In this chapter, we have demonstrated that these regions are also internally coherent—that is, they are a single, internally interdependent economy or economic region.

Perhaps the most salient example of variation among local economies is found in the examination of local business cycles in Chapter 5. The existence of distinct local business cycles also demonstrates the internal integration of local economic regions, but, more important, it undermines the image of "national trends" that are independent of local realities.

NOTES

1. Data reprinted with permission from Ledebur, L. C., & Barnes, W. R. (1993). *All in it together, cities, suburbs and local economic regions.* Washington, DC: National League of Cities.

2. Voith examined the population and real per capita income growth over a 30-year period (1960-1990) in 28 metropolitan areas within the northeast and north central regions. He found that throughout the 1970s and 1980s, city and suburban population and income growth were positively correlated, indicating that a complementary relationship exists. Voith, R. (1992, September/October). City and suburban growth: Substitutes or complements? In *Business review* (pp. 21-33). Philadelphia: Federal Reserve Bank of Philadelphia.

3. Savitch, H. V., Collins, D., Sanders, D., & Markham, J. P. (1993, November). Ties that bind: Central cities, suburbs and the new metropolitan region. *Economic Development Quarterly, 7*(4), 341-357. Savitch and colleagues also identified a relationship between prices of office space in central cities and suburb and edge cities. They found a correlation of 0.86, indicating that the increasing or decreasing price of "downtown office space can impact suburban economies" (p. 346).

4. Adams, C., et al. (1996, March). Flight from blight and metropolitan suburbanization revisited. *Urban Affairs Review, 31*(4).

5. Hill, E. W., Wolman, H. L., & Ford, C. C., III. (1995, November). Can suburbs survive without their central cities?: Examining the suburban dependency hypothesis. *Urban Affairs Review, 31*(2), 149. See also the response by H. V. Savitch, "Straw men, red herrings, . . . and suburban dependence" and the authors' response, both in *Urban Affairs Review, 31*(2), 147-183.

6. Ibid. p. 151.

7. See for example: Dreier, P. (1995, July/August). Making the case for cities. *Challenge,* 29-37. Todd Swanstrom [in Swanstrom, T. (1996, May). Ideas matter: Reflections on the new regionalism. *Cityscape, 2,* 5-21] indicates that the greatest weakness of the new regionalism debate is that it is "trapped in the terms of the liberal/conservative debate between more government or freer markets" (p. 15) and speaks of the "extremes of the new regionalism debate" as the choice between a new layer of regional governments or governmental fragmentation (p. 13). Although it is difficult to identify or even to characterize the terms of the new regionalist debate (most participants are surprised to find themselves so categorized), regional government or, where it already exists, even more regional government, is not the solution that falls naturally from the recognition that the regional economy is a collective or common good. Rather, it is the growing awareness of the need for governance, not government restructuring—collaboration among new jurisdictions, not metropolitan government—that is giving rise to regionalism. The issue is not liberal or conservative; it is not about more or less government but about how existing governments can work together to tend the economic commons.

8. Downs, A. (1994). *New visions for metropolitan America* (p. 58). Washington, DC/Cambridge, MA: The Brookings Institution/Lincoln Institute of Land Policy.

9. Cisneros, H. G. (Ed.). (1993). *Interwoven destinies: Cities and the nation* (p. 24). New York: Norton.

CHAPTER 5

BUSINESS CYCLES AND LOCAL ECONOMIES

Economic theory and federal policy appear to assume (a) that there is a single national economy, (b) that there are uniform regional responses to federal macroeconomic policies, and (c) that variations in regional business cycles and their responses to national policies are not the concern of national policy and can safely be ignored. Policies targeted to regions or industries are dismissed as "structural" and considered by "mainstream" economists not to be an appropriate means of promoting cyclical stabilization.

If there is a single national economy, however, and if regions respond uniformly to federal macroeconomic policies, then the business cycles experienced by regions should conform to the pattern of the national business cycle and, hence, to one another. The national business cycle and regional fluctuations, measured by rates of change in cyclical indicators, should be one and the same or similar enough that they would not matter.

Business cycles have been the nemesis of industrial economies. Karl Marx predicted the ultimate collapse of capitalism through successive and increasingly severe cycles of boom and bust. The promise of Keynesian economics was its potential to moderate these recurrent cycles of depression and expansion. Mainstream economics since the 1930s and federal economic policy since the 1960s have focused on means to control cyclical instability and promote stable, long-term national economic growth.

The youthful art of macroeconomics tends not to deal with differences in the timing and severity of business cycles across regions or with differences in the responsiveness of regional cycles to federal fiscal and monetary policies. The only interest in regional differences has focused on the possibility of using regions that tend to lead the national cycle as a means of forecasting national trends.

The existence and persistence of substantial regional variations in sensitivity to business cycles, however, is well documented.[1] This evidence, largely ignored by the mainstream of the economics profession, discredits the assumptions of uniformity in regional business cycles and regional responses to federal stabilization efforts. Furthermore, it points to the inadequacy of the nationalist economic paradigm and suggests the shape of the needed new view of economic reality.

REGIONAL BUSINESS CYCLES

Inquiry into the region as a focus of business-cycle analysis appears to have begun with the work of Rutledge Vining at the University of Virginia. As World War II drew to a close, Vining began to explore the thesis that a national sum or average used to measure economic fluctuations is actually an "estimate of a frequency distribution of corresponding measurements of the component parts of the nation."[2]

Vining's starting point was the work of the respected economist Joseph Schumpeter, who a decade before had argued that aggregate analysis not only does not tell the whole tale but also necessarily obliterates the main point of the tale.[3] Building from an analogy of physics, Vining argued,

The "wave" that is observed in national averages and sums appears as something different if not entirely apart from the movements exhibited by the underlying economic "particles" or units. The motion of numbers of these units is always counter to the direction of the national "wave," and a national "turning point" appears merely as a shift in the direction of movement of the preponderance of the "particles." . . . The turning point occurs when a particular direction of movement ceases to be "counter" and assumes its role of ascendancy. Aggregate analysis does not show these adjustment processes of economic evolution.[4]

According to this analysis, Keynesian macroeconomics, along with the very rudimentary tools of regional economics and data, is "fraught with shortcomings." Indeed, Vining found that regions experience different rates of economic change over business cycles, and that the national rate of change in income is only a measure of the central tendency of a frequency distribution of regional rates of change.[5]

In the intervening years, analyses using improved analytical tools and data sets have supported and expanded Vining's thesis and conclusions. In general, these studies concluded that the timing and severity of business cycles varied across regions, and that these differences were explained only in part by sectoral composition. Other factors of importance in explaining differences in regional cyclical behavior were growth differentials, intraindustry variations, and specific regional factors.

Variations in the timing and duration of business cycles across states, regions, and labor market areas reveal themselves in a variety of ways. Some areas have not experienced cyclical behavior during national business cycles.[6] Others have local up and down cycles during periods of uninterrupted national economic growth. In general, the timing of turning points in state and market area cycles has not corresponded to those of national cycles.

Studies of the cyclical performance of metropolitan areas have found that some metropolitan areas tend to lead the national business cycle, whereas others lag behind the national cycle or correspond to the pattern of the national cycle.[7] Further documentation of the diversity of regional cyclical behavior was provided by Victor and Vernez in their study of three major business cycles between 1960 and 1976.[8] Their study, the researchers argued, "again demonstrates the need for efficient targeting of countercyclical relief based on local rather than national indicators."[9]

Also differing among regions is the severity of business cycles.[10] Wide variations are found in the responses of states, metropolitan areas, and manufacturing sectors to patterns of national economic change.[11] Furthermore, the sensitivity of urban areas to national patterns and the importance of local factors in determining unemployment differ greatly.[12] Given these variations in the timing and severity of business cycles, the capacity of macroeconomic policy alone to respond to persistent disparities in the regional impacts of recessions has been questioned.[13]

The preponderance and the consistency of the evidence on regional business cycles appear to undermine substantially two of the primary assumptions supporting the national orientation of federal policies: the existence of a single national economy and the likelihood of uniform regional responses to federal policies triggered by national statistical averages.

REGIONAL MARKETS AND REGIONAL CYCLES

Despite significant differences in regional cyclical behavior, economic theory provides a sophisticated justification for excluding regional considerations from the domain of national policy.

A central tenet of neoclassical economics holds that free market forces will tend to reduce and, ideally, eliminate economic disparities among subnational areas. The assumption is that there is a tendency in market economies toward an equilibrium of factors of production (including cost and availability of materials, labor, capital equipment, and financial capital) across regions. Market signals will direct resources to the areas of greatest productivity and highest return. Persistent differences in rates of return to factors of production, by this argument, occur because of "frictions" in market adjustments, "rigidities" in markets, and interventions that distort market signals.

Ostensibly, this was the logic underlying the dismantling of much of the corpus of federal regional and urban policy in the early 1980s. At the time, the Reagan administration argued that its role was to "level the playing field" rather than to pursue policies that would distort the operation of the marketplace by intervening to affect the welfare of particular places.

Surely, the role of markets is essential in intraregional, interregional, and, therefore, national productivity, economic efficiency, and growth. The prevailing thesis, however, suggests that, over time, the performance of economic regions in the United States will converge, with differences reflecting only short-term adjustments. If markets lead to convergence, however, it would be expected that regions would become more similar in their cyclical behavior over time. Regional variations in business cycle behavior would be self-correcting, and therefore any differentials could be ignored by the framers of macroeconomic policy.

There does not, however, appear to be any trend toward convergence in regional responsiveness to the national business cycle. Over the six classic business cycles between 1948 and 1977, the relative severity of regional responses to the business cycle has been substantially different. Furthermore, although there has been no consistent widening or narrowing of regional differences, regional responses have been somewhat more varied in recent years.[14] Even after adjusting for secular trends in employment, there is still no indication of convergence.

SOURCES OF REGIONAL VARIATIONS

In light of the evidence clearly documenting the persistence of regional variations in business cycles, economists and economic geographers have sought to explain the sources of these regional differences. These explanations range from differences in growth rates, regional industry composition, and industry performance to regional competitive advantages. In almost every case, the evidence points to the importance of region-specific factors as well as to the particular mix of industries in a region.

Regional Cycles and Secular Growth

Regional sensitivity to business cycles and regional rates of economic growth are interrelated.[15] Regions with higher rates of growth and per capita income experience smaller declines in downswings and greater expansions in upswings than more slowly growing areas.[16] This evidence also supports the argument that regions with greater cyclical sensitivity experience lower growth rates and vice versa. Wide cyclical swings create uncertainty that dampens secular growth rates.

Three explanations of the relationship between cyclical sensitivity and growth have been advanced. First, growth occurs because of more efficient, low-cost firms that can compete more effectively in cyclical downturns than less efficient firms. Second, the geographical distribution of innovation and technological progress is weighted in favor of some regions.[17] Third, more prosperous regions will experience population growth that increases local demand for commodities, housing, and public infrastructure and services. These increases in demand, in turn, permit a region to better weather cyclical downswings. Furthermore, investor expectations of future growth support levels of private investment when cyclical swings are viewed as temporary.[18]

Regions and Sectors

The extent of variation in regional cyclical behavior should not be particularly surprising. The expectation of conformity of regional cycles with the statistical artifact of the "national cycle," an average of all regions, would be reasonable only if (a) the industrial composition of all regions was essentially the same, (b) the cyclical experience within each industry was the same in every region, and (c) all local factors affecting economic performance were equal or irrelevant.

Industry structure consistently has been identified as an important factor in determining a region's sensitivity to the business cycle.[19] Some industries, such as durable manufacturing, are particularly sensitive to business cycles. Often, mature and slow-growing industries never fully recover from cyclical downturns in expansion phases of the national and regional cycle. Sectors facing heavy international competition may also fail to recover fully from recessions. In other words, regions specializing in cyclically sensitive or mature, slow-growth industries—together with regions exposed to aggressive foreign competition—clearly experience cycles of greater severity than those specializing in less sensitive sectors.[20] An increasing concentration of employment in real estate, finance, insurance, and services tends to cushion a region during downturns, whereas a higher concentration of manufacturing appears to exacerbate their severity.[21]

Differences in the cyclical experiences of regions, however, are greater than anything that can be explained by industry mix alone.[22] The industrial composition of a region—although it is one of the factors responsible for differences in cyclical behavior—is not the only factor. Often, in fact, it is not even the most important factor in determining a region's performance over the business cycle.[23] Other factors specific to regions also play an important role.

At its most naive, the industry-mix thesis assumes that each individual industry is spatially homogeneous in its cyclical behavior. If there are regional differences in cyclical behavior within industries, however, regional industry composition per se will be a poor predictor of regional cyclical behavior.

The cyclical behavior of industrial sectors is not homogeneous. There are marked differences among regions in cyclical behavior within individual industries. Furthermore, differences in regional industry composition or mix may be diminishing in importance as causes of interregional variations in cyclical behavior. Gaining in importance are intraindustry variations across regions.[24] Since World War II, a process of structural convergence has been occurring with the industry mixes of major U.S. regions becoming more similar rather than more distinctively specialized.[25] Despite this structural convergence, regional cyclical experiences have not become more similar.[26]

Regional Sensitivity to Monetary and Fiscal Policy

Studies over a period of almost 45 years have documented pronounced differences in regional cyclical behavior—differences attributable to both industry composition and region-specific characteristics. Given the extent of these regional differences, it would be expected that federal monetary and fiscal policy would have differential impacts.

The impact of federal policy on regional performance was tested in a brief flurry of studies focusing on the monetarist critique of the application of Keynesian analysis to regional economies.[27] Although the focus of these studies was on methodological differences, their findings contribute to the knowledge of regional impacts of federal policies. The following two conclusions, supported in varying degrees, have emerged:

1. Both fiscal and monetary policy influence regional economic performance.
2. There are pronounced differences in the regional impacts of both monetary and regional policy.

Such findings led some researchers to promote the efficacy of regionally oriented stabilization policies. For example, a 1965 study concluded the following:

> Employment fluctuations vary regionally and . . . regionally pinpointed stabilization measures increase the efficiency of stabilization policy. But this emphasis on regional aspects does not imply as a corollary that the responsibility for stabilization should be left to regional governments.
>
> The fact that the federal government is the most efficient unit for undertaking stabilization measures and for bearing their costs does not eliminate a concern with regional factors, nor does it follow that state (and local) governments have no role to play in stabilization policy. Indeed, through proper inter-level cooperation, a more efficient regional orientation can be secured than would be possible by reliance on purely federal policies.[28]

Eight years later, another study reached very similar conclusions:

> Recognition of the interaction between regional and industrial distribution is essential in the design of both general and sectoral economic policies. If the nation is viewed as a network of related regional markets, regionally oriented stabilization policies may be appropriate. To date, fiscal policy has implicitly assumed that increases in aggregated demand are spatially uniform in their impact. However, the evidence of non-uniform industry declines across regions implies that the location of the initial expenditure increase is important. . . . The federal government is clearly the most efficient unit for undertaking stabilization policies and for bearing their costs. Nevertheless, a regionally oriented macro-economic policy does not have to be purely federal.[29]

This promising strand of research appears to have flourished only briefly and to have waned in the early 1980s, yielding to the national orientation of macroeconomic research and policy.[30]

REGIONAL CYCLES AND NATIONAL POLICY

Regional business cycles are a pervasive reality. There are persistent differences in the timing and severity of cyclical change across regions. What is commonly referred to as the "national business cycle" is simply the average of changes occurring in the nation's regions. The timing of federal responses to the national business cycle, therefore, will affect regions at different phases of local business cycles of varying severity. Uniform regional responses to federal policies are highly improbable.

Given the documented differences in regional cyclical behavior, these uniform national policies will be procyclical in some regions and countercyclical in others and, thus, will fail to maximize the growth potential of some regions. Part of the monetarist's critique of fiscal policy as a macroeconomic tool centers on the lag entailed in undertaking discretionary policy measures. According to this argument, by the time the need for a change in federal tax or spending policy to either slow or stimulate the national economy is recognized, and appropriate fiscal measures are legislated, the economy is likely to be in a different phase of the business cycle, and the measure is as likely to be procyclical as countercyclical. This critique is easily extended to uniform national fiscal and monetary measures directed to the national statistical cycle.

The persistence of significant differences among regions in cyclical experience indicates that regions will not respond uniformly to nationally oriented countercyclical policies. Because of these persistent differences in the timing, as well as the severity, of regional cycles, the geographical focus and impacts of countercyclical measures will be an important factor in determining their effectiveness. Regionally discriminating policies have the potential to reduce the magnitude of cyclical swings and promote secular economic growth.

The federal government is the most efficient level of government for implementing countercyclical policies in both their national and regional dimensions. Other units of government, however, can play a role as well. If governments within the regional economies can find ways to work in concert, their regionally aggregate investment and tax policies, coordinated with federal measures, should be an important component of regional countercyclical efforts.

Variations among regional business cycles are anomalous in the context of the nationalist economic paradigm. They bring together the findings of Chapter 3 regarding variations among metropolitan regions and those of Chapter 4 regarding the internal coherence of those regions. The accumulated evidence suggests the inability of the nationalist paradigm to account for economic reality, underscoring the need for a different lens for viewing the economic world.

NOTES

1. For reviews of some of this literature, see the following: Strong, J. S. (1983). Regional variations in industrial performance. *Regional Studies, 17,* 429-430; Vaughan, R. J. (1976). *Public works as a countercyclical device: A review of the issues.* Santa Monica, CA: RAND.

2. Vining, R. (1945, July). Regional variation in cyclical fluctuation viewed as a frequency distribution. *Econometrica, 13,* 183.

3. "Price levels, totals of deposits and expenditures, net losses, and so on becoming the variables to be handled, the whole economic system in depression acquires the features of a losing concern whereas it is the essence of the process that the individual elements of those aggregates are differentially affected." Schumpeter, J. (1934). *Business cycles* (p. 134 and Note 1, p. 153). New York: McGraw-Hill.

4. Vining, R. (1946). The region as a concept in business-cycle analysis. *Econometrica, 14,* 202.

5. Vining, R. (1946). The region as a concept in business-cycle analysis. *Econometrica, 14,* 212.

6. Vernez, G., Vaughan, R., Burright, B., & Coleman, S. (1977). *Regional cycles and employment effects of public works investment.* Santa Monica, CA: RAND. This study was undertaken to investigate the feasibility of public works investments as a countercyclical device. It examined the cyclical performance of 149 labor market areas (LMAs) in 47 states and the District of Columbia between 1960 and 1975, a period covering three business cycles using monthly employment data. A labor market area is defined by the Bureau of Labor Statistics as an economically integrated geographic unit with a central city or cities and the surrounding territory within commuting distance. In most cases, labor market areas are coincident with MSAs. Also see the following: Connaughton, J. E., & Madsen, R. A. (1986, Spring). Recession and recovery: A state and regional analysis. *Review of Regional Studies, 16*(2), 76-86.

7. Casetti, E., King, L., & Jeffrey, D. (1971). Structural imbalance in the U.S. urban-economic system, 1960-65. *Geographical Analysis, 3,* 239-255; Casetti, E., King, L., & Jeffrey, D. (1972, September/October). Cyclical fluctuation in unemployment levels in U.S. metropolitan areas. *Tijdschrift Voor Econ. en Soc. Geografie*; and Casetti, E., King, L., & Jeffrey, D. (1969). Economic impulses in a regional system of cities. A study of spatial interaction. *Regional Studies, 3,* 213-218.

8. Victor, R. B., & Vernez, G. (1981, March). *Employment cycles in local labor markets.* Santa Monica, CA: RAND.

9. Victor, R. B., & Vernez, G. (1981, March). *Employment cycles in local labor markets* (p. 19). Santa Monica, CA: RAND.

10. The most detailed study of regional cyclical performance in the United States examined the duration, amplitude, and severity of business cycles in metropolitan areas and labor market areas, in addition to the differences in time discussed previously

[Vernez, G., Vaughan, R., Burright, B., & Coleman, S. (1977). *Regional cycles and employment effects of public works investment*. Santa Monica, CA: RAND]. It concluded that (a) there are large variations between states and labor market areas in the characteristics of their employment cycles in response to national economic fluctuations; (b) similarly, the duration, amplitude, and severity of employment cycles vary considerably across states and labor market areas; and (c) cycle severity—the rate of cyclical unemployment—across labor market areas varied from a low of 5% to a high of 19%. The average LMA experienced a cycle severity of 4.4% to 5.2%.

11. Borts, G. H. (1969). Regional cycles of manufacturing employment in the United States, 1914-1953. *Journal of the American Statistical Association, 55*, 151-211; Phillips, B. D. (1972). A note on the spatial distribution of unemployment by occupation in 1968. *Journal of Regional Science, 12*(2).

12. Casetti, E., King, L., & Jeffrey, D. (1971). Cyclical fluctuation in unemployment levels in U.S. metropolitan areas. *Tijdschrift Voor Econ. en Soc. Geografie.*

13. Syron, R. F. (1978, November/December). Regional experience during business cycles—Are we becoming more or less alike? *New England Economic Review*, 2-34. Differences in the timing and severity of business cycles among regions are not unique to the United States. Significant differences in cyclical performance among subnational areas have been found for Great Britain, The Netherlands, Northern Ireland, and Canadian provinces. One study of regional cycles in Great Britain concluded that the usefulness of national unemployment rates in stabilization policy is seriously impaired by the tendency of regions to lead or lag the national rate and differences in regional sensitivity to national cyclical movements (Brechling, 1976, p. 19). Another study, also of unemployment in Great Britain, concluded that, to increase the efficacy of unemployment policies, governments should both undertake policies to ensure steady employment growth and "take differential policy measures to ensure that regions suffer or benefit more equally as national conditions deteriorate or improve" (Thirlwall, 1966, p. 217). Brechling, F. (1976). Trends and cycles in British regional unemployment. *Oxford Economic Papers, 19*, 1-21; Bassett, K., & Haggett, P. (1981). Toward short-term forecasting for cyclic behavior in a regional system of cities. In M. Chisholm, A. E. Frey, & P. Haggett (Eds.), *Regional forecasting*. London: Butterworth; Hewings, G. J. D. (1978). The trade-off between aggregate economic efficiency and interregional equity: Some recent empirical evidence. *Economic Geography, 54*, 254-263; Thirlwall, A. P. (1966). Regional unemployment as a cyclical phenomenon. *Scottish Journal of Political Economy, 13*, 205-215; Van Duijn, J. J. (1975). The cyclical sensitivity to unemployment of Dutch provinces, 1950-1972. *Regional Science and Urban Economics, 5*, 357-374; Black, W., & Slattery, D. G. (1975). Regional and national variations in employment and unemployment—Northern Ireland: A case study. *Scottish Journal of Political Science, 22*, 195-205; King, L. J., & Clark, G. L. (1978). Regional unemployment patterns and the spatial dimensions of macro-economic policy: The Canadian experience 1966-1975. *Regional Studies, 12*, 283-296; Clark, C. L. Factors influencing the space-time lags of regional economic adjustment. *Annals of Regional Science, 15*, 1-14.

14. Clark, C. L. Factors influencing the space-time lags of regional economic adjustment. *Annals of Regional Science, 15*, 29; Syron, R. F. (1978, November/December). Regional experience during business cycles: Are we becoming more or less alike? *New England Economic Review*, 29; Strong, J. S. (1983). Regional variations in industrial performance. *Regional Studies, 17*, 434.

15. Borts, G. H. (1969). Regional cycles of manufacturing employment in the United States, 1914-1953. *Journal of the American Statistical Association, 55*, 151-211; Engerman, S. (1965). Regional aspects of stabilization policy. In R. A. Musgrave (Ed.), *Essays in fiscal*

federalism (pp. 27-29). Washington, DC: The Brookings Institution; Vaughan, R. (1976). *Public works as a countercyclical device: A review of the issues.* Santa Monica, CA: RAND.

16. Victor, R. B., & Vernez, G. (1981, March). *Employment cycles in local labor markets* (p. 18). Santa Monica, CA: RAND.

17. Richardson, H. W. (1969). *Regional economics* (p. 277). New York: Praeger.

18. Engerman, S. (1965). Regional aspects of stabilization policy. In R. A. Musgrave (Ed.), *Essays in fiscal federalism* (p. 28). Washington, DC: The Brookings Institution.

19. Borts, G. H. (1969). Regional cycles of manufacturing employment in the United States, 1914-1953. *Journal of the American Statistical Association, 55,* 151-211; Browne, L. E. (1978, November/December). Regional industry mix and the business cycle. *New England Economic Review;* Engerman, S. (1965). Regional aspects of stabilization policy. In R. A. Musgrave (Ed.), *Essays in fiscal federalism.* Washington, DC: The Brookings Institution; Kasarda, J. D., & Irwin, M. D. (1988, September). [National business cycles and community competition for jobs]. Research sponsored by the National Science Foundation; Strong, J. S. (1983). Regional variations in industrial performance. *Regional Studies, 17,* 434.

20. A corollary to the thesis that industrial mix determines regional cyclical sensitivity is the view that industrial diversification will buffer subnational economies from the vicissitudes of the business cycle. At one level, the diversification argument simply states that increasing the share of the regional economic base in less cyclically sensitive export industries will reduce cyclical magnitudes in the regional economy. At this level, the diversification thesis is a subset of the industry composition argument. At a second level, diversification argues for "import substitution," the substitution of locally produced commodities and services for those imported into the region. Import substitution, however, is likely to increase the cyclical sensitivity of a regional economy because of the capacity to pass on internal declines in income to other regions by reducing imports. See Engerman, S. (1965). Regional aspects of stabilization policy. In R. A. Musgrave (Ed.), *Essays in fiscal federalism* (p. 26). Washington, DC: The Brookings Institution.

21. Victor, R. B., & Vernez, G. (1981, March). *Employment cycles in local labor markets* (pp. 17-19). Santa Monica, CA: RAND.

22. Borts, G. H. (1969). Regional cycles of manufacturing employment in the United States, 1914-1953. *Journal of the American Statistical Association, 55,* 151-211.

23. Browne, L. E. (1978, November/December). Regional industry mix and the business cycle. *New England Economic Review,* 48. The most detailed critique of the industrial composition argument found some support for the role of industry mix in the cyclical experiences of regions, but the importance of this industry effect was over the downward side of the regional business cycle. During the expansion phase of cycles, no significant relationship was identified (Engerman, S. [1965]). Regional aspects of stabilization policy. In R. A. Musgrave [Ed.], *Essays in fiscal federalism.* Washington, DC: The Brookings Institution).

24. These conclusions were drawn by Vaughan [Vaughan, R. (1976). *Public works as a countercyclical device: A review of the issues.* Santa Monica, CA: RAND] from a review of existing evidence on regional business cycles and causes of differential cyclical behavior.

25. Strong, J. S. (1983). Regional variations in industrial performance. *Regional Studies, 17,* 431. Research in Great Britain also found that intraindustry differences across regions are more important in explaining regional cyclical behavior than regional industry mix. See Harris, C. P., & Thirlwall, A. P. (1968). *Interregional variations in cyclical sensitivity to unemployment in the United Kingdom, 1949-1964* (Vol. 30, pp. 55-66). Oxford, UK: Oxford University Institute of Economic and Statistics.

26. Strong, J. S. (1983). Regional variations in industrial performance. *Regional Studies,* 17, 431, 434; Borts, G. H. (1969). Regional cycles of manufacturing employment in the United States, 1914-1953. *Journal of the American Statistical Association, 55,* 201.

27. Beare, J. B. (1976). A monetarist model of regional business cycles. *Journal of Regional Science, 16*(1), 57-64; Garrison, C. B., & Chang, H. S. (1979). The effect of monetary and fiscal policies on regional business cycles. *International Regional Science Review, 4*(2), 167-180; Andersen, L. C., & Jordan, J. L. (1968). Monetary and fiscal actions: A test of their relative importance in economic stabilization. *Federal Reserve Bank of St. Louis Review, 50,* 11-21; Andersen, L. C., & Carlson, K. M. (1970). A monetarist model for economic stabilization. *Federal Reserve Bank of St. Louis Review, 52,* 7-25; Mathur, V. K., & Stein, S. (1980). Regional impact of monetary and fiscal policy: An investigation into the reduced form approach. *Journal of Regional Science, 20*(3), 343-351; Garrison, C. B., & Kort, J. R. (1983). Regional impact of monetary and fiscal policy: A comment. *Journal of Regional Science, 23*(2), 249-261; Mathur, V. J., & Stein, S. H. (1983). Regional impact of monetary and fiscal policy: A reply. *Journal of Regional Science, 23*(2).

28. Engerman, S. (1965). Regional aspects of stabilization policy. In R. A. Musgrave (Ed.), *Essays in fiscal federalism* (pp. 53, 56). Washington, DC: The Brookings Institution.

29. Strong, J. S. (1983). Regional variations in industrial performance. *Regional Studies,* 17, 443.

30. Regional and urban economics tend to be viewed as distant cousins of suspicious origins within the economics profession, and their contributions are largely ignored in the corpus of economic theory and national policy making. Other studies of note in this literature include the following: Borts, G. H. (1969). Regional cycles of manufacturing employment in the United States, 1914-1953. *Journal of the American Statistical Association, 55,* 151-211; Browne, L. E. (1978, November/December). Regional industry mix and the business cycle. *New England Economic Review,* 35-53; Syron, R. F. (1978, November/December). Regional experience during business cycles: Are we becoming more or less alike? *New England Economic Review,* 25-34; Vaughan, R. (1976). *Public works as a counter-cyclical device: A review of the issues.* Santa Monica, CA: RAND; Vernez, G., Vaughan, R., Burright, B., & Coleman, S. (1977). *Regional cycles and employment effects of public works investment.* Santa Monica, CA: RAND.

CHAPTER 6

ECONOMIC FEDERALISM
AND THE NEW POLITICAL
ECONOMY

This chapter explores two paradigms. The first is a new economic paradigm; the second is a paradigm of political economy, or the melding of economics and politics from which economic policy emerges. The economic paradigm changes the way we think about the economy and economies; it shifts the framework and metaphors of economic thinking. The political economy paradigm incorporates the new economic paradigm and changes the way we think about politics, governance, and economic policy.

Changing our economic paradigm, our "truths," is difficult and can be painful. The origins of change are in the fatigue of the prevailing paradigm—in this case the nationalist view of the economy, a view that identifies the "nation as the economy" or the "economy as the nation."

What follows, therefore, is conceptual—the beginnings of a process of developing a new understanding of a changed and changing economic reality. These beginnings are alternative images and conceptual frameworks. They cannot necessarily be proven. Their test is in their capacity to explain what is truly happening and in their utility in guiding effective economic policy.

ECONOMIC REGION

The cornerstone of this new economic paradigm is the local economic region (LER). From the LER, looking outward, we see a national system of regional economies and, beyond, a "global economy." Turning, looking inward, is the urban system that gives internal coherence and integration to the LER. The local economic region provides a framework for demonstrating the interdependent relationships between a region and its smallest economic unit and among local, regional, "national," and global economies as well.[1]

In the minds of many people, the term *local economy* is synonymous with *local political jurisdiction*—for example, the (local) economy of the city of Austin. Thus, the term has, over the years, developed a meaning that is incorrect and that misleads policymakers. Local governments are not local economies. The real (functional) local economy is the local economic region.

Also distracting are different uses of the term *region.* In some cases, region is used to refer to multistate territories such as the Census Bureau's Northeast, Midwest, South East, and South Central regions. These, however, are classifications of convention and convenience, reflecting contiguity and propinquity more than any other logic of organization or function.

A use of region with greater precision, and certainly greater utility, is the concept of the "nodal region" found in the literature of regional economics and geography. This is where our concept of the local economic region is grounded—in the line of reasoning that uses "nodes" or "nodal" to refer to urban centers or the "urban centeredness" of regions.

Regional economic theory helps us identify three primary characteristics of the nodal region.

First, the economic region is functionally integrated and highly interdependent. Functional integration results from vertical, horizontal, and complementary economic relationships within the region.[2] Vertical relationships exist when firms in the region serve as markets for products produced within the region (forward linkages) or as suppliers of materials and other inputs for firms producing in the region (backward linkages). Horizontal relationships exist when economic activities compete for the same resources, inputs, and markets within the region. Complementary relationships exist when increases (or decreases) in one economic activity in the region result in the growth (or decline) of other regional activities through forward and backward linkages.[3]

Second, the economic region is economically oriented toward one or more dominant urban centers or nodes. In nodal regions, flows of population, goods, services, communication links, traffic patterns, and so on are polarized "towards and from one or two dominant centers."[4]

Third, the urban centers are primary sources of developmental innovations that drive the economic growth of the region. Two types of interactions characterize functional regions. Noninnovative spatial interactions include the daily routines of commuting and shopping to maintain the system. Innovative spatial interactions involve diffusion of technology and ideas that help an area to grow and develop.[5]

The economic region, the local economy of the new economic paradigm, is rooted in these three core concepts. The following define the economic region:

1. The economic region is the basic building block of the economy.
2. It is a single, integrated, and interdependent regional economy overlaid by the political jurisdictions of cities, suburbs, and surrounding non-metropolitan areas.
3. Economic regions are centered around, or polarized to, one or more urban or metropolitan areas. The fulcrum of the local economic region is the metropolitan area, not "the city" or any governmental jurisdiction.
4. These metropolitan centers are the sources of new ideas, new technologies, and innovations that drive economic growth and development within the region and throughout the national system of economic regions. It is the metropolitan centers of the economic regions, rather than the core cities alone, that are the source of creativity and innovation.
5. Economic regions, therefore, are an essential resource in the economic growth process. They are generators or engines of growth in the national economic system.

Some observers may object to the idea that economic regions are metropolitan centered and that nonmetropolitan cities and towns, as well as rural areas, are a part of these regions. However, in a detailed study of U.S. urban systems, Edgar Dunn concludes that there are no "rural" areas; all regions, he argues, are effectively linked to some urban core in today's modern society.[6] If this is the case, the geography of the United States will be completely partitioned by wall-to-wall economic regions, with all areas included within some metropolitan-centered economic region.

If, however, there is found some area that is not economically integrated into a metropolitan-centered economic region, it will likely be underdeveloped and poor. It may be politically difficult for nonmetropolitan areas to accept that they are under the economic sway of metropolitan areas, but it is economically unrealistic to think otherwise. Residents of these areas will be better served by working to strengthen their economic linkages to the metropolitan centers of their regions rather than by building political barriers through claims of economic autonomy and independence.

Thus, an economic region is larger than its metropolitan center to the extent that this center is economically integrated with the nonmetropolitan areas of its periphery. Because the economic importance of economic regions is only now beginning to be understood, however, we do not yet have the means to delineate them unambiguously. This problem also stems from the way data are collected in the United States—by political jurisdictions, cities, counties, and states.

Jurisdictionally based data are insufficient to the task of revealing and understanding the economic relationships among places where economic boundaries are ambiguous. Another problem in "mapping out" the nation's local economic regions may be that some areas, through different economic functions, may be economically interdependent with more than one economic region.

Neal R. Peirce and colleagues use the term *citistates* in discussing regions. In their recent book of the same name, they suggest that

> The very ambiguity of the definition of citistate may tell us a lot. The citistate is the most dynamic form of human settlement today. We are just beginning to sense its full, latent power. But it may be years before we grasp its limits, geographic or political.[7]

To this we would add, "or economic."

Multiple political jurisdictions are nestled within the geography of the urban system of the economic region. These jurisdictions are governmental entities created to pursue the political and public service goals of their residents. They are not economies, however. The economic region is the functional economy. Thus, a region can be said to include one economy and many jurisdictions. Economic and political boundaries do not coincide.

Because policy and government are organized by jurisdictions, and not by local economic regions, the salient question is the following: How do we govern and nurture a metropolitan economic system in which political boundaries are not congruent with the functional regional economy?

THE "NATIONAL ECONOMY"

Metropolitan-centered economic regions are the basic economic units—the foundation of the national economy. These are not the semiautonomous, independent economies of Jane Jacobs's city regions.[8] They are open systems. Each economic region is linked to other economic regions. The economic future of any single region, therefore, is interdependent and interwoven with the economic destinies of many other regions but not equally with all of them. This complex, interdependent system of metropolitan-centered economic regions is the true national economy.

Robert Reich has argued that the national economy is "the region of the global economy demarcated by the nation's political boundaries."[9] It can also be argued that the real national economy is not contained by national boundaries or constructs of government.[10] Rather, it spills across the borders of the United States and beyond the political jurisdiction of the federal government. Seattle-Vancouver, Detroit-Windsor, San Diego-Tijuana, and El Paso-Juarez, for example, are single regional economies divided by national boundaries. In each case, neither nation nor national government can lay singular claim to the divided economic region. These regional economies are part of the economic systems of both nations, their national economies.

The national economy is thus an open system, just as its constituent economic regions are open. Through the economic regions, the national economic system is linked and, therefore, interdependent with other national economies around the world. National economic independence, therefore, is an illusion—a myth that does not serve us well. The reality of the emerging global economy is the interdependence of national economic systems and LERs. It is a reality that must be reflected in economic policy at all levels.

INTERDEPENDENCE OF REGIONS

Economic regions are not self-sufficient. Because they are open systems, their productivity and performance depend on flows of products, services, materials, resources, technology, innovations, ideas, and information from the national economic system as well as the global system of national economies. These flows permit economic regions to specialize, to some degree, in economic activities in which they have competitive advantages.

By specializing in regional competitive advantages, economic regions achieve higher levels of economic performance. As a result, the productivity of the system as a whole, both the national economy and the global economy, increases.

Economic specialization and regional performance and productivity, however, are highly dependent on the strength and efficiency of the system of linkages that connect economic regions. Regions with weaker linkages to the larger economic system invariably will be less specialized and less developed and will perform more poorly as a result.

Regional specialization and the increasing reliance on linkages to other local economies directly translate into growing interdependence in the wider system of economic regions. Understanding the national economy as a national system—and grasping the essential interdependence of economic regions—helps us respond to the perennial question of subnational economic policy in the United States: Why should one region care what happens in another? Framed in another way, why should national economic policy focus on economic regions rather than retaining a singular orientation toward what is perceived to be the undifferentiated national economy?

In other words, why should Seattle, with its international orientation to the Pacific Rim, care what happens to Cleveland, and why should federal policymakers care as long as Seattle is doing well? The answer is that Seattle and Cleveland are economic regions in the national economic system. Even if Seattle's economy is not directly dependent on what happens in Cleveland, Seattle's economic performance is deeply and directly affected by the national system and the region's linkage to the national economy.

Consider, by analogy, a spider's web. A point at one edge of the web will not be directly linked by any single strand to a point on the opposite edge. If a strand is plucked at the first point, however, the disturbance will radiate throughout the web because of a radial network of strands. Ultimately, vibrations will be experienced at every point on the web, including points on the opposite edge. Seattle, as part of the national economy, will be similarly affected by the economic fortunes and misfortunes of Cleveland.

This is the nature of complex systems. Regional economies are interdependent in such a way that significant economic changes in any region will result in related changes in others.

ADAPTIVE SYSTEM

The national economy is an adaptive system. The competitiveness of the national economic system and each of its economic regions depends on the capacity to evolve, adapt, and respond to the changing requirements of the international and national economies as well as to emerging opportunities created by new technologies, new innovations, and changing competitive advantages. The long-term performance of the national economic system will depend on its ability to adapt.

The systemic goal, therefore, is not to resist or inhibit the process of adaptive change in economic regions but to promote and facilitate it. The industrial restructuring that has occurred in many regional economies during the past 25 years represents an adaptation of the system. Although the adverse consequences to people and places are of deep concern, the goal for the national economy must be to facilitate these processes of industrial adaptation and to propel the regions on a path of renewed economic growth and development that maximizes the productivity and the outcomes of the national system.

Illusions of economic autonomy can also drive adverse patterns of economic change. Jurisdictions, for example, often make decisions driven primarily by localized interests—decisions that result in minimal improvements in productivity and competitiveness, often at great cost. Political autonomy permits these decisions. A sense of economic independence and autonomy thus abets and encourages policies that are zero-sum or negative-sum for the region.

Corporations, too, can make counterproductive decisions for the economic regions and the national system. Just because a relocation decision represents a gain for one firm, for example, does not necessarily mean it will benefit the economy as a whole. Indeed, such decisions, made solely within the context of the interests of a single enterprise, may drive spirals of change within economic regions that, in net, diminish regional productivity and growth. This suggests that the link between a firm's autonomy and economic efficiency—a connection that underscores our faith in competitive markets—may not hold in all cases.

MAPS OF ECONOMIC REALITY

A map of the U.S. interstate highway system presents a visual image of physical linkages among regions. Although this image is useful in considering the new economic paradigm, however, it is an oversimplification that risks the casual assumption that the economic regions are most closely linked to those most proximate to them. Clearly, in many cases this will be true, although advances in transportation and communications technologies are reducing the time and cost benefits of propinquity and proximity.

An alternative, and perhaps more accurate, visual image for the new paradigm can be drawn from airline routes and the electronic links of a national communication system. This alternative image is of a national system less defined by the tyranny of distance and space. This image reflects the fact that, in the national economic system, economic regions may be more closely linked and more intensely interdependent with faraway regions than with those close by.

Patterns of regional interdependence transcend national boundaries as well. In an open national economic system, economic regions can be closely linked and interdependent with the economic regions of other national economies: the West Coast economic regions of the United States and the economic regions of the Pacific Rim; East Coast economic regions and those of the European common market; and, especially since approval of the North American Free Trade Agreement, U.S. economic regions throughout the nation and those of Canada and Mexico.

A NEW ECONOMIC PARADIGM

This new image of economic reality is certainly more complex and less symmetrical than the long-dominant image of one national economy. It is also more difficult to organize through a neat set of concepts. It is this more complex reality, however, that should be the crucible of economic policy. Key and distinctive elements of this new paradigm of economic reality include the following:

1. The regional paradigm recognizes that economic regions, not state or local political jurisdictions, are the basic building blocks of the national economy.
2. It focuses on the internal integration and interdependence of economic regions, and it recognizes the essential interdependence of the parts of the urban economic system.
3. It understands that the national economy is the national system of economic regions, not the political jurisdiction of the federal government.
4. It establishes a clear focus on the interdependence of economic regions and the critical importance of linkages to the performance of the national economic system, the true national economy.
5. It recognizes that the national economy spills across the jurisdictional boundaries of the federal government.
6. It emphasizes that the national economy, the system of economic regions, is an open system inextricably linked to other national economic systems in the global economy.
7. It recognizes that governmental boundaries can and do have significant economic effects.

THE DILEMMA OF TWO FEDERALISMS

The United States has a federal system of government consisting of three tiers of government, a political federalism.[11] The U.S. economy consists of two tiers: the metropolitan-centered local economic region and the national economic system (the national economy). We call this "economic federalism."

The U.S. political economy can be described as a system shaped by two federalisms: one economic and one political. Both contain multiple and often conflicting loyalties and investments. Of immediate importance here, however, are the relations between the two.

The two federalisms are not spatially congruent. Nowhere are the lines of economic federalism and political federalism aligned. Each set of boundaries—one governmental and the other economic—has its own history, its own dynamics, and its own justifications.

Neither set of boundaries is, necessarily, more "real" or more "justified" than the other. Political boundaries around cities, counties, states, and the nation are the result of geography, war, treaty, political maneuver, legislative action, and historical happenstance. From an economic perspective, these lines are completely arbitrary.

Local economies' boundaries, however, are defined by historical patterns of development, natural and man-made competitive advantages, and past and currently prevailing market forces and investments. From a governmental perspective, these economic boundaries may themselves appear arbitrary.

The mutual arbitrariness and noncongruence of these two sets of boundaries cannot easily be eliminated. Also, for the purposes of policy making, this noncongruence constitutes a major obstacle that cannot be ignored. The fact that our political and economic boundaries are so different means that our political system is prevented from seeing the economy clearly. This noncongruence can also muffle or distort the ability of economic interests, especially those of a metropolitan region taken as a whole, to voice their concerns. An unfortunate result is that these conditions can severely inhibit effective policy making.

Ignoring the noncongruence of political and economic boundaries promotes frustration and political instability. Goals cannot be met, and conflict over the economy cannot appropriately be framed or joined or both. Indeed, during the past 200 years, noncongruence between the two federalisms may have resulted in greater freedom of action in the economic sector precisely because any clear view of the real economy from the political side has been thwarted. The same situation, however, may have also resulted in a greater degree of frustration, conflict, and acute sensitivity on the part of economic interests and actors because our policy tools have been inappropriate to the real nature of the economy.

The central government and state and local governments mistakenly believe that they each have jurisdiction over "an economy," when in fact they do not. Political jurisdictions are unable to identify their real economic interests because no jurisdiction is congruent with the economic region, and thus no jurisdiction is compelled to look at the region as a whole.

These situations—along with other conditions—allow narrower interests to become the dominant voices within each jurisdiction and at each "level" of government.[12] Asserting the primacy of the economic region will not thwart narrow-interest domination. It will, however, remove one critical support for such domination.

The chief adverse consequence of the noncongruence of political and economic boundaries, as we have noted, is that the making and implementation of relevant and effective policy are hampered. Some accommodation between the real economy—the metropolitan-centered economic region—and governmental structures is a prerequisite for good policy. Thus, this barrier must be overcome or at least circumvented to create a framework, a new political economy, within which effective policy can be made and carried out.

THE CAULDRON OF POLITICAL ECONOMY

Separate paradigms of both economics and politics are not the framework in which to develop policies and programs. The true framework of policy development and implementation is the political economy, the mortar of economics as affected by the pestle of politics.

To discuss economic policy, therefore, one must deal with the economy and with government, and one must deal with them fused to-

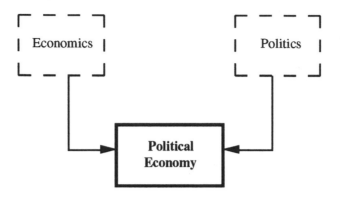

Figure 6.1. Fusion of Politics and Economics

gether, which is, after all, the way they actually exist (Figure 6.1). Thus far, we have discussed "economy" abstracted from this setting to offer a new understanding of its spatial structure and operations. Now, ideas about economics must be intertwined with ideas about government and politics as the political economy of both the metropolitan-centered economic region and the nation. To set the stage for this task, it is useful to think about political economy in the context of the nationalist paradigm of economy and political federalism (Figure 6.2).

Central to the nationalist political economy is the belief that (a) there is a single, homogeneous national economy and (b) this economy and the political jurisdiction of the United States are closely aligned. In this paradigm, the real economy of goods and services and the monetary economy (monetary policy, fiscal policy, regulatory standards, and income redistribution) are also aligned. This assumption of alignment of the economy and central government underlies a political economy in which the federal government makes policy for the national economy. If this assumption of alignment is incorrect, then our political economy is distorted and our policy making is likely flawed. Good policy does not flow from flawed frameworks.

According to this framework of the national political economy, other levels in the federal system are not viewed as having, and do not assume, any role in national economic policy making. Policy involvement flows downward through the vertical political system—never

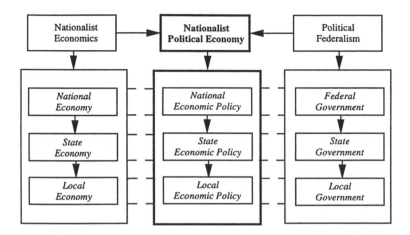

Figure 6.2. Fallacy of Alignment

upward or interactively. There is no forum within the prevailing nationalist political economy in which localities are consistently heard and legitimately "at the table" (except as "special interests").

A discordant subplot of the prevailing political economy is the perception that state and local government jurisdictions also encompass economies within the national economy. This thesis is accepted as true despite its dissonance with the prevailing nationalist economic paradigm. One explanation for the misperception is that state and local representatives simply replicate the national error—that the political entity constitutes an economy. Nonetheless, from the national viewpoint these subnational entities are, economically, special interests and thus not legitimate participants in national economic policy making.

In this subnational dimension of the political economy, jurisdictions and subnational economies are aligned; they are one and the same. States administer economic development policies and programs as if the states were economies. The federal government administers programs to assist state governments in their economic development efforts. Local governments also pursue economic development strategies for their jurisdictions that reflect the assumption that they are coherent economies with economically meaningful borders. Also, to bring the process full circle, both federal and state governments provide assistance for these local economic development programs.

Although the shape of political economy changes slowly over time, or more quickly in response to pressing events, its central compass, the nexus of economic nationalism and political federalism, provides a sustaining orientation for economic policy. The main reason behind the longevity and persistence of this central tendency of economic policy is apparent in Figure 6.2. The underlying paradigm of political economy is symmetrical and, in its symmetry, quite simple. The systems are in alignment. The mental maps and images—together with the metaphors of the economy, the polity, and the political economy—are clear and compelling in providing a compass for public policy.

Effective policy making requires that government be able to see the economy clearly. Under the nationalist economic paradigm, the line of sight is seemingly unobstructed—from the national government to the national economy. The two entities encompass the same space, and only one government is involved. This sense of clarity of vision, however, is an illusion, a distortion of economic reality that impedes the national quest for global competitiveness.

Local governments also believe they have an unobstructed view of economic reality, from their political jurisdictions to the local economy. States, too, see the economy and the polity as one and the same. This perception of clarity, however, is a distortion of economic reality that diminishes, rather than enhances, the productivity of each economic region and, hence, of the national economic system.

TOWARD A NEW POLITICAL ECONOMY:
THE U.S. COMMON MARKET

Reality is never as simple nor as symmetrical as the orderliness we impose on it and through which we understand it. The reality of two federalisms, one economic and one political, is asymmetrical and significantly more complicated than the reigning nationalist economic paradigm, both in images and metaphors and as a context for the formulation of economic policy. It is this more complex reality, however, in which the United States must shape economic policies for global competitiveness.

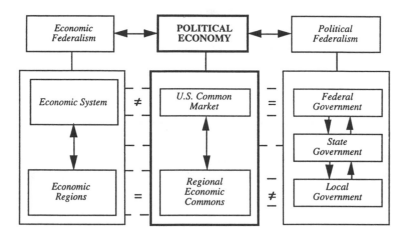

Figure 6.3. Political Economy of Two Federalisms

To build a new political economy, we must start by recognizing that the United States has two federalisms—rather than the single political federalism of the political economy of nationalism—and, equally important, that these two federalisms are neither symmetrical nor congruent (Figure 6.3). As the federal government turns to the economy, it confronts not a single economy neatly circumscribed by national geographic boundaries but rather an economic system of interdependent local economic regions. A visual image of such an economic system is presented in Figure 6.4, in which hexagons represent local economies (local economic regions); the system's boundaries are depicted by the dark line circumscribing the local economies.

Economic systems spill across national boundaries in politically untidy ways. Economic systems are interwoven with other economic systems in ways that cannot be captured in two-dimensional models of bilateral trading relationships. The economic system is, most importantly, internally heterogeneous.

The boundaries of the federal jurisdiction do not circumscribe a single economy but rather a network of economic regions. Thus, the federal government confronts an economic system not congruent with the federal jurisdiction or the economy of money and regulation (Figure 6.5). The portion of the economy within, or mostly within, the

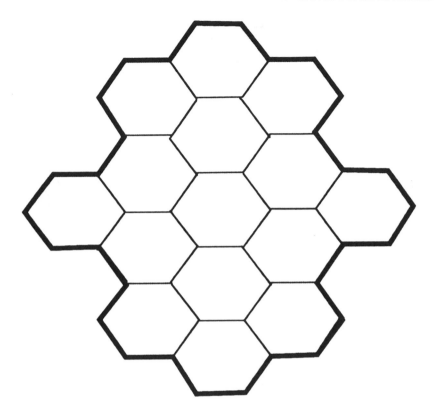

Figure 6.4. Economic System of Local Economic Regions

jurisdictional boundaries of the nation is a common market of economic regions. The U.S. common market, therefore, is a system of interdependent economic regions within national boundaries and is subject to the economic policies of the federal government. The geographical boundaries of the common market, the federal jurisdiction, and the monetary and regulatory economy are aligned.

This is the real political economy of the United States. In this political economy, it is clearly recognized that the U.S. common market is not a single economy, nor is it the whole of the economic system. Effective economic policy developed within this political economy, therefore, will attempt to nurture the whole of the economic system while seeking to strengthen its building blocks, the economic regions. Clearly, eco-

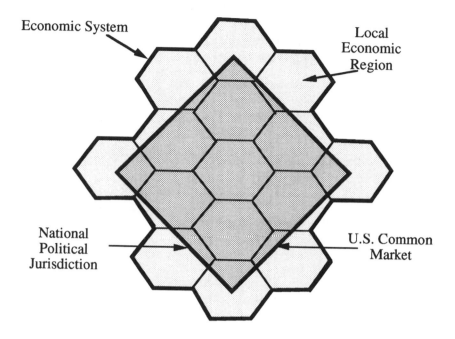

Figure 6.5. U.S. Common Market

nomic policy holds greatest sway within the U.S. common market, but it never loses sight of linkages to economic regions beyond the political jurisdictions of the federal government.

OVERVIEW

Chapter 1 previewed a regional paradigm composed of a three-tiered economic system and a three-tiered political economy (Figure 6.6).

The focus of this chapter has been on the national political economy characterized by two noncongruent federalisms, one political and one economic. Chapters 7 and 8 explore the other two tiers of the macro-economic system and their political economies. Chapter 7 focuses on the local economic region and the political economy of the regional economic commons. In turn, Chapter 8 examines the global economy

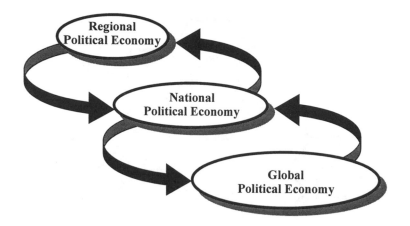

Figure 6.6. Interdependent Economic Systems

through the lens of the regional paradigm. In Chapters 6 through 8, we thus present a sweeping overview of the dimensions of this new paradigm.

NOTES

1. See Barnes, W. R., & Ledebur, L. C. (1991). Toward a new political economy of metropolitan regions. *Government and Policy, 9*, 127-141; Ledebur, L. C., & Barnes, W. R. (1992). *City distress, metropolitan disparities and economic growth.* Washington, DC: National League of Cities; Ledebur, L. C., & Barnes, W. R. (1994). *All in it together: Cities, suburbs and local economic regions.* Washington, DC: National League of Cities; Barnes, W. R., & Ledebur, L. C. (1994). *Local economies: The U.S. common market of local economic regions.* Washington, DC: National League of Cities.

2. Hoover, E. M. (1975). *An introduction to regional economics.* New York: Knopf; Richardson, H. W. (1969). *Regional economics.* New York: Praeger.

3. For a more detailed discussion of these relationships, see Henderson, W. L., & Ledebur, L. C. (1972). *Urban economics: Processes and problems.* New York: John Wiley.

4. Richardson, H. W. (1969). *Regional economics* (p. 227). New York: Praeger.

5. Berry, B. J. L., Conkling, E. C., & Ray, D. M. (1976). *The geography of economic systems.* Englewood Cliffs, NJ: Prentice Hall. These authors also discuss political, or adminis-trative, regions defined by political boundaries following a geographic separation of power. They note that "boundary changes for such administrative areas generally do not keep pace with the changing geography of economic systems" (p. 248). Brian Berry refers to his version of the economic area as the "daily urban system" (DUS). The DUS delineates

the urban area's functional boundaries using the daily commuting fields that include the residential or housing market areas as they relate to particular employment or job market clusters. Originally developed by C. A. Doxiadis in 1967, the DUSs defined by Berry, in conjunction with the Bureau of Economic Analysis, are based on evidence of commuting patterns and related information such as newspaper circulation, telephone traffic, and road networks. These spatial units completely disaggregate the United States and have a high degree of closure with respect to housing and job markets. They are multinodal with a high degree of linkages and interactions within them. Berry, B. (1973). *Growth centers in the American urban system* (Vols. 1-2). Cambridge, MA: Ballinger.

6. On the basis of this conceptual image of urban structures, Edgar S. Dunn, Jr. [Dunn, E. S., Jr. (1980). *The development of the U.S. urban system*. Baltimore, MD: Johns Hopkins University Press] concludes that the best available representations of regions that exist in reality are Brian Berry's daily urban system/Bureau of Economic Analysis "economic area" constructs.

7. Peirce, N. R., Johnson, C., & Hall, J. S. (1993). *Citistates: How urban America can prosper in a competitive world* (p. 13). Washington, DC: Seven Locks Press.

8. Jacobs, J. (1984). *Cities and the wealth of nations: Principles of economic life* (p. 180). New York: Random House.

9. Reich, R. (1991). *The work of nations: Preparing ourselves for the 21st century capitalism* (p. 244). New York: Knopf.

10. Here we confront a problem of concept and language. Do we use the term *national economy* in reference to the national economic system that spills across national boundaries or to that portion of this economic system within national geographical boundaries. The concept of a U.S. common market of economic regions refers to the political economy of the network of economic regions contained within the national boundaries. The term *national economy* could, therefore, be used to refer to this network of economic regions within the national jurisdiction, and an alternative term, such as *U.S. economy*, could be used to refer to the larger national economic system that transcends national boundaries. Here, we have adopted the convention of referring to the national economy as the national economic system rather than the elements of that system within national boundaries.

11. Jones, B. (1989, September). Why weakness is a strength: Some thoughts on the current state of urban analysis. *Urban Affairs Quarterly, 25,* 31-34. Although the U.S. Constitution attributes sovereignty to only the central and state governments, most commentators refer to three "levels" of government in the American federal system. This usage is followed here because it is more realistic and more relevant for policy purposes. It is also more correct in reflecting the current condition and historical development of American governments. In effect, the U.S. Constitution alone does not constitute the foundation of American federalism. It must be read in conjunction with the 50 separate state constitutions, which in turn establish the "local" level in the three-tiered federal system.

12. See, for instance, Molotch, H., & Logan, J. R. (1987). *Urban fortunes: The political economy of place.* Berkeley: University of California Press.

THE REGIONAL
ECONOMIC COMMONS

Debate about the relationships of cities and suburbs has too often degenerated into an unproductive "either/or" controversy serving narrow political and social interests. Much of metropolitan politics and public discourse are at an apparent dead end on this issue.

Current infatuation with the image of economic autonomy and fragmentation has bolstered claims of suburban economic, social, and cultural independence from their greater regions. Such claims echo those of an earlier age and very different circumstances of the "independence" of central cities from their regions and, hence, their suburbs. Both views, current and past, deny the commonality and economic interdependence of jurisdictions within the economic region. Each is dangerously incorrect.

Cities and suburbs are political jurisdictions astride a single regional economy. Jurisdictions within a regional economy are, therefore, economically interdependent. The nature and dimensions of this interdependence vary from place to place, but interdependence is nonetheless an economic reality. Denial of this essential reality fosters the seeds of the spatial suicide occurring in many of our nation's urban areas.

The damage wrought by this image of economic fragmentation and autonomy must be repaired. To do so requires a clear vision of the integral nature of political economy of economic regions. Achieving this vision again necessitates separating economies from polities, and economics from politics, before subsequently rejoining them as a new political economy of regions.

THE DILEMMA OF LOCAL "STATISM"

The modified nationalist or local statist paradigm (see this volume, pp. 75-76), with its simple and appealing sense of symmetry and orderliness, provides the foundation of the image of jurisdictional autonomy and independence. Political boundaries and economies appear congruent. Each rectangle or square in Figure 7.1 represents a separate political jurisdiction, and the larger rectangle represents the central city of the region. Through the lens of modified nationalism, each jurisdiction is a separate economy competing with other jurisdictional economies within (and beyond) their region.

In this view, each jurisdiction is, for all practical purposes, an autonomous unit of government bound only by the state authorities or restrictions under which it operates. Each has its own tax base. Each makes specific policies and investment decisions and provides a specific mix of services, amenities, and taxes ostensibly preferred by their citizens.

In the eyes of "public choice" advocates, this pattern of governmental "fragmentation" permits all citizens of the region to choose the jurisdiction that provides their preferred "mix" of services, amenities, and tax obligations. This choice permits individuals and families to maximize their locational advantage subject only to the constraints of income or wealth.

Critics of the public choice perspective counter that the outcome of this system of government is sharp disparities in tax bases among jurisdictions across urban areas with disturbing differences in the quality of life, services, and amenities and relative tax burdens experienced by citizens across the region. In other words, this model of locational

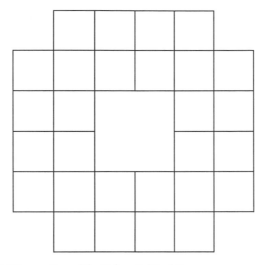

Figure 7.1. Statist Perspective of the Urban Political Economy

choice underlies the pattern of income and, therefore, racial and social segregation that prevails in most metropolitan areas.

The jurisdictionally focused blinders of this perspective inevitably encourage competitive actions to promote local tax bases without consideration of any consequences to other jurisdictions or the regional economy. Incentives for interjurisdictional cooperation are limited, by and large, to economies of scale in the provision of some services, significant spillover effects, responses to external threats posed by actions of state and federal governments, or, in rare cases, federal or state programs to promote some form of cooperative action.

The Achilles' heel of the statist perspective at the local level is that using this paradigm makes it exceedingly difficult for jurisdictionally based politicians and citizenry to correctly perceive the true nature of the economy held in common and the costs of misperceiving or ignoring its interdependence. It also makes coordinating actions among two or more jurisdictions extraordinarily difficult.

THE REGION AS AN ECONOMY

The regional paradigm provides a very different view of reality. To understand this perspective, the reader is asked to mentally and visually roll up the latticework of jurisdictions overlaying the regional economy and, for the present, set this rolled lattice aside. Doing so permits examination of the regional economy unimpeded by the vision-narrowing blinders of the jurisdictional framework.

In attempting to understand the economy without jurisdictions, it is important to keep sight of the grounding of the economy in all its facets in social contexts that must be viable if the economic components are to be successful, just as the economy must succeed if these social environments are to be viable. The concept of the urban system—the social, political, institutional, and economic system of the region—captures this broader context.

Figure 7.2 presents a visual representation of the regional economy. It assumes the hexagonal form introduced in Chapter 6. At the spatial center of this economy is the historical central business district around which many or most regional economies grew. Arrayed across the plane of this economy are activity centers: residential villages (neighborhoods), economic villages (clusters of retail, commercial, and service activities), and industrial villages (clusters of manufacturing, warehousing, shipping, wholesale, etc.).

Dispersed throughout the region are a set of edge cities that include the central business districts of other central cities in the region (e.g., Dearborn and Pontiac in the greater Detroit region) and the recent edge cities. In *Edge City*, Joel Garreau achieved a glimpse of the organizational form of the metropolitan economic region.[1] Garreau's insight was that, in suburbs, there is no single urban center or hub. Rather, the development pattern in suburban areas is characterized by multiple clusters of development and economic activity. In the jargon of urbanologists, the suburbs and metropolitan area as a whole are polycentric (multiple-centered or "multinodal") rather than monocentric (organized around a single central core as in the traditional view of the city).

This economic region is a complex, integrated, highly interdependent system—a complex network of clusters of economic, social, institutional, and residential activities. It consists of multiple clusters or nodes of development and economic activity that are linked and interdependent. Each cluster of specialized activities, including residential, is part of an economically integrated whole. This linked, interdependent system of development nodes is the organizational form of the economic region. Neither edge cities nor central cities, nor any other

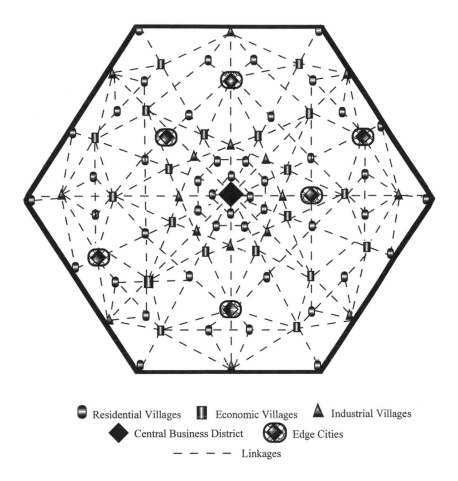

Residential Villages   Economic Villages   Industrial Villages

Central Business District   Edge Cities

— — — — Linkages

Figure 7.2. Local Economic Region

parts of the local economic region, are economically autonomous and independent.

The view of the economic region as a complex, interdependent nodal system emphasizes the critical role of the linkages or functional relationships (dashed lines in Figure 7.1) that create and sustain the system. Edgar Dunn argues that urban systems can be described by classifying actors, places, and physical linkages and by detailing these functional relationships among these actors, places, and linkages.[2]

The most obvious of these linkages are the regional networks of physical infrastructure, such as the transportation and communications networks that connect the urban system. The greater the efficiency of these linkages—that is, the lower the time and money cost of sustaining economic and social activity patterns—the greater the productivity and interdependence of the system.

Although physical infrastructure is the linkage that is most easily visualized and that, perhaps, serves best to communicate the image of functional relationships among nodes, other nonphysical linkages are no less important. Among these are service delivery systems such as health, education, and safety; regional financial systems; intergovernmental linkages; and human resource systems.

The economic region is also the regional labor market. The productivity of this local economy will be directly affected by the efficiency of the linkages between workers and jobs. The more regional this system—the more it transcends jurisdictions and proximity—the more productive the local economy will be, with the labor market becoming less exclusionary.

In this intraurban system, the central business district of the central city historically was, and often remains, a crucial center of functionally specialized economic, social, and institutional activity. It also is true that these central areas are most often the locus of governance functions as well as more specialized business, cultural, health, entertainment, and recreational activities that serve a wider regional audience.

Clearly, the degree of influence or dominance of this central node varies among metropolitan regional economies. In some, particularly economic regions that are international centers, the role of the central business is greater. In others, the role and regional influence of the central business district are weaker. Regardless, in an interdependent regional economy, core cities play an important role that has not been supplanted by edge cities.

The isolated, point-in-time image of Figure 7.2 obscures two key characteristics of the regional economy. First, this complex, interdependent regional economic system is not static or unchanging. It is a dynamic, evolving system continuously adapting to changing local imperatives and external demands. The long-term economic vitality, productivity, and social health of each regional economic system will depend on its ability to change, to adapt, and to reconfigure its spatial form and patterns of economic activity.

Second, the regional economic system is neither self-contained nor autonomous. An essential characteristic of the economic region is that it is an open system, inextricably linked to other economic regions, and thus its welfare is interdependent with other economic regions. Just as the destinies of components within an urban system are "interwoven,"[3] the fates and futures of economic regions are interdependent because of the openness of these urban systems.

THE REGIONAL ECONOMIC COMMONS

In the regional paradigm, the economic (the economic region) and the political (local jurisdictions) must be re-fused into the political economy of the region. This is shown in Figure 7.3 in which the lattice of political jurisdictions (Figure 7.1) overlays the regional economy (Figure 7.2). In this political economy, the boundaries of the economy and boundaries of the polities are not the same. No single jurisdiction lays claim to the whole of the economy or, as represented here, incorporates a major portion of the economic base of the region.

Figure 7.3 is a generic representation of the map of almost any urban region. Economic and residential clusters or nodes are found, in greater or lesser degree, within the boundaries of each jurisdiction. In many cases, these activity nodes fall completely within a jurisdiction. In others, however, these clusters spill across jurisdictional boundaries and are part of what is perceived to be the economic base of two or more jurisdictions. Linkages and functional relationships of the economic system are not constrained by jurisdictional boundaries, although jurisdictions have powers to promote or diminish the efficiency of these systems linkages and relationships through their investment decisions and regulations.

Policies and uncoordinated investment patterns that are driven by narrowly construed jurisdictional interests can erode economic linkages and functional relationships, diminishing the productivity of the economic system as a whole. This is apparent in urban transportation systems that are not regionwide but are provided by jurisdictions that

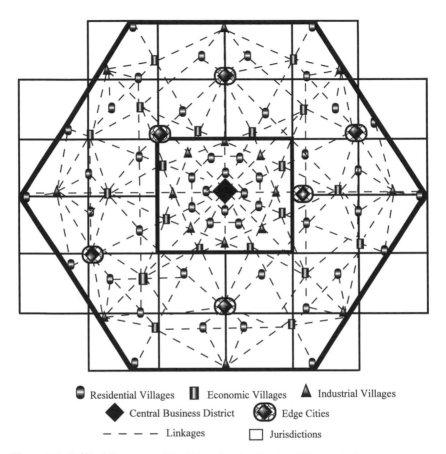

● Residential Villages ▮ Economic Villages ▲ Industrial Villages
◆ Central Business District ⊕ Edge Cities
– – – – Linkages ☐ Jurisdictions

Figure 7.3. Political Economy of the Urban Region: Regional Economic Commons

often make it difficult and costly to move throughout the system. This issue is also important in relation to significant differences in regulations, environmental standards, and so on and in the quality and availability of basic utilities and services such as public education, public safety, and open housing.

David Rusk, in *Cities Without Suburbs*, argues that cities that are "elastic" or able to expand their jurisdictional boundaries through annexation fare far better than those with tax bases limited by inelastic boundaries.[4] This argument can be expanded in the context of the regional paradigm. Increasing the proportion of the regional economy

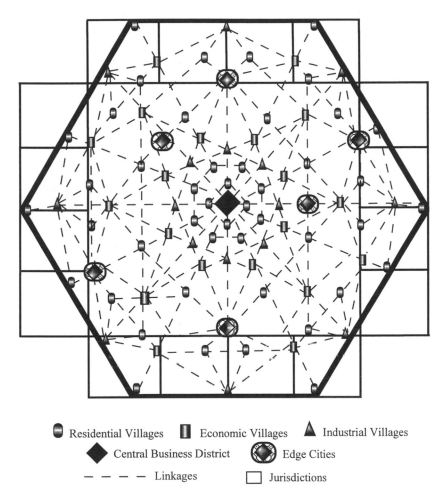

● Residential Villages   ▮ Economic Villages   ▲ Industrial Villages

◆ Central Business District   ◉ Edge Cities

— — — — Linkages   ☐ Jurisdictions

Figure 7.4. Regional Economic Commons With Dominant Core City

and tax base within a single political jurisdiction both reduces the barrier effects of jurisdictional boundaries on linkages and functional relationships and increases the possibility of making policies and investment decisions that promote the productivity of the regional economy. Visually this is seen in Figure 7.4, which differs from Figure 7.3 in the share of the regional economy falling within the jurisdictional boundaries of the central city.[5]

Under the regional paradigm, the regional economy is held in common. The blurred boundaries of the regional economy and the sharply defined jurisdictional boundaries of the polities are not aligned—not congruent. In almost every place, the regional economy is latticed with many local political jurisdictions and many stakeholders.

A single local economy with multiple jurisdictional stakeholders raises fundamental questions of local economic governance and local economic policy making. It poses the dilemma of the commons. Who tends the regional economic commons when no single jurisdiction holds exclusive rights? Who nurtures the economy held in common?

This regional economy is a "collective good" in the classic sense. If one unit of government were congruent with the local economy, the benefits (and costs) of economic policies would accrue within that jurisdiction. In the economy held in common, however, public action to promote the economy creates benefits that cannot be entirely captured by the investing jurisdiction. The local government undertaking the action bears the cost but cannot capture all the benefits. Absent collective action, therefore, local governments inevitably will underinvest in the nurturing of the regional economy because of these spillover effects.

The defining characteristic of the regional economic commons, therefore, is interdependence. This stands in stark contrast to the misleading sense of economic autonomy derived from the statist paradigm.

The interdependence of the commons has two key dimensions, one economic and one political. Economic interdependence arises from the very nature of the economic region as a complex interdependent system. What affects any element or dimension of the system affects the whole and, directly or indirectly, each of the parts.

Political interdependence arises from the nature of the economic commons. The regional economy is held in common by many jurisdictions and many stakeholders. Households live in one jurisdiction, shop in another, and work in another. The welfare of each jurisdiction within the region, therefore, will be affected by the actions of other jurisdictions through their effects on the economic commons as a whole.

There is also vertical economic and political interdependence in the political economy of the regional paradigm. Each local economic commons is an interdependent part of the national common market. Actions that affect the common market affect each economic commons. In turn, local actions that affect a single economic commons directly or indirectly impact all others in the common market. Local governments, therefore, have an important stake in what federal and state governments do that affect the common market and any one of its economic commons. In turn, federal and state governments have important stakes in local actions that affect the regional economic commons. Recognition of these vertical and horizontal dimensions of economic and political interdependence in the economy is critical to effective policy making at each level of government in our federal system.

Recognition that the region is a single economy, an economic commons, overlaid with multiple political jurisdictions provides a radically different perspective for policy and policy making. If interdependence is a signature economic characteristic of the region, collaboration becomes a key to success in economic governance of the commons. In other words, if economic interdependence is a fact, collaboration is a defining political challenge. A region's success in responding to this challenge may be a critical determinant of its economic future.

GROWING THE REGIONAL ECONOMY

This view of the regional economy, absent the lattice of jurisdictions, provides an important perspective on "growing the economy." If the singular goal of a region is to enhance the economic productivity of the economic region, in what should we invest? What would be the appropriate level of investment in each area or function? How would we prioritize these investments?

Some answers to this difficult question are obvious. Investments would be made in continuously upgrading the quality of the regional workforce and other important factors of production as well as sustaining and enhancing critical economic development infrastructure. Strategic investments would be made in ensuring that essential economic functions are performed well and efficiently while avoiding wasteful duplication of capacity and facilities. Investments would be made in social service networks of efficient scales. Investments would be targeted to developing clusters of specialized economic activities and residential patterns and transportation systems that facilitate the economic functions of these clusters while minimizing time, money, and disamenity costs of transporting workers, materials, and products.

This perspective of the economy is also useful in identifying improbable and potentially inefficient investments and investment patterns. If there is a focus on the growth and productivity of the regional economy,

It is highly unlikely that investments would be made that create sharp disparities in the quality of schools across the regional economy that adversely impact the quality of the regional workforce.

It is highly unlikely that investments would be made in spatially fragmented transportation systems.

It is highly unlikely that infrastructure investments would be made that facilitate economically inefficient and wasteful accelerated turnover in housing and building stock.

It is highly unlikely that letting parts of the economy's critical infrastructure fall into disrepair would be viewed as effective stewardship of the economy's scarce competitive resources.

It is highly unlikely that investments would be undertaken that simply move economic activity around within the region without contributing to regional productivity.

It is highly unlikely that segregating households by income class or race would be viewed as efficient or rational economic development policies.

This litany of productive and unproductive investments makes clear that a line of sight to the regional economy from each jurisdiction both clarifies needed actions and highlights the economic irrationality of many current policies and investment patterns.

DISPARITIES IN THE COMMONS

This perspective on the regional commons brings focus to a critical issue at the heart of many of our nation's metropolitan areas: Do sharp economic disparities within the region matter?

Clearly, there are important social and political reasons why persistent disparities within the regional commons matter. Widening disparities between cities and suburbs will inevitably undermine social cohesion and political legitimacy.[6]

Do economic disparities, however, matter to the productivity and growth of the economic commons? Do these disparities matter to suburbs and more prosperous segments of the economic commons? Do

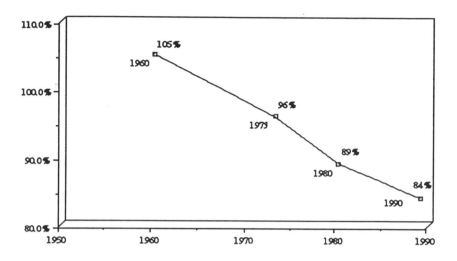

Figure 7.5. Central City and Suburban Income Disparities, 1960-1990. Reprinted with permission from Ledebur, L. C., & Barnes, W. R. (1992, September). *City distress, metropolitan disparities and economic growth: Combined revised edition* (p. 2). Washington, DC: National League of Cities.

thcy matter to other regional economic commons in the U.S. common market?

The core issue is whether the regional economic commons or the U.S. common market can prosper if large segments of each are impaired and unable to contribute to system maintenance, productivity, and growth. It is unfortunate that the debate surrounding this issue is almost always in terms of cities and suburbs. Political dynamics of multijurisdictional metropolitan areas frame the debate in these terms. The fact that almost all social and economic data are collected and reported by jurisdictions compounds the inclination to view the issue in city-suburban terms. It is very difficult to address the issue of the effects of disparities on economic performance outside this restrictive city-suburb framework.

A substantial decline in the economic welfare of cities relative to their suburbs has been occurring since at least 1960 (Figure 7.5). In 1960, per capita income was slightly greater in central cities of metropolitan areas than in their suburbs. By 1973, per capita income in central cities had fallen to 96% of their suburbs. By 1980, this ratio had fallen to 89%. This decline in the economic welfare of cities relative to suburbs continued

Percent MSA
Employment Growth

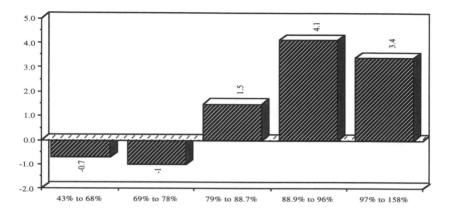

Figure 7.6. City and Suburban Economic Disparities and Metropolitan Growth. Calculated for the 85 largest MSAs with central cities in 1990. Per capita income ratios are for 1989. Employment growth rates are for the period January 1988 to August 1991. Bars represent average rates of employment growth for 85 cities divided into quintiles (17 cities per quintile).
SOURCE: U.S. Bureau of the Census, 1990 Census; Bureau of Labor Statistics, Washington, D.C.

in the 1980s. By 1990, per capita income in central cities was down to 84% of suburban income.

A great deal of attention has focused on the polarization of incomes in the United States in the past decade. The *1994 Economic Report of the President* devoted a prominent subsection to this topic, calling it a "threat to the social fabric."[7] The data in these figures suggest a spatial dimension to this issue. Quality of life is directly related to geographical places and the economic vitality of these places. The long-term trend toward increasing disparities between per capita incomes in cities and their suburbs means that the issue of economic inequality must also be addressed in this spatial context.

Evidence of a relationship between disparities and metropolitan economic performance is strong, although additional research remains to be done. Figure 7.6 examines the average rate of employment growth from 1988 to 1991 for the largest 85 metropolitan areas, categorized by the per capita income disparities between cities and their suburbs.

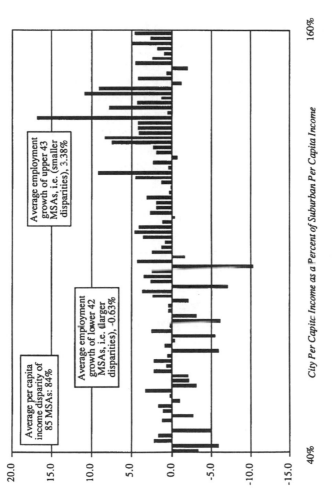

Average per capita
income disparity of
85 MSAs: 84%

Average employment
growth of lower 42
MSAs, i.e. (larger
disparities), -0.63%

Average employment
growth of upper 43
MSAs, i.e. (smaller
disparities), 3.38%

City Per Capita Income as a Percent of Suburban Per Capita Income

Figure 7.7. City and Suburban Disparities and Employment Growth.

SOURCE: U.S. Bureau of the Census, 1990 Census; Bureau of Labor Statistics, Washington, D.C.

NOTE: Calculated for the 85 largest MSAs with central cities in 1990. Per capita income ratios are for 1989. Employment growth rates are for the period January 1988 to August 1991. Bars represent average rates of employment growth for 85 cities divided into quintiles (17 cities per quintile).

As Figure 7.6 shows, there is a direct relationship between city-suburban disparities and rates of employment growth. Metropolitan areas with lower disparities tend to have higher rates of employment growth. Those with lower employment gains tend to have higher disparities and those with higher employment gains tend to have lower disparities.

The tendency for lower disparities and higher rates of employment growth to go together is apparent in Figure 7.7, which presents rates of employment growth for individual metropolitan areas ranked by levels of disparities in per capita income. There are exceptions to this generalization. A few metropolitan areas with relatively higher income disparities experienced relatively high rates of employment growth. These exceptions may be due to the size of these cities relative to their suburbs, unique characteristics of their economic bases and other particular local or regional characteristics, or they may simply be exceptions to the general trend.

Other research has corroborated this finding.[8] In a study for the National Bureau of Economic Research, Roland Benabou explored the relationships among education, income distribution, and growth at the local level.[9] On the basis of data for 14 metropolitan areas, he identified "a strong positive relationship between the lack of economic segregation, measured by the ratio of central city to suburban mean incomes, and the area's growth in both per capita income (1969-1989) and total employment (1973-1988)."[10] This he terms a "striking stylized fact" that does not permit any inference about causality but does suggest that city-suburban stratification and metropolitan economic performance are interdependent processes that require systematic economic analysis to identify the underlying mechanisms.

Benabou develops what he refers to as an integrated framework for analyzing the determinants of inequality and growth and starting to untangle the lines of causality.[11] On the basis of this analysis, he concludes that "minor differences in educational technologies, preferences, or wealth can lead to a high degree of income stratification," and that "stratification makes inequality in education and income more persistent across generations." Furthermore, he finds that "the polarization of urban areas resulting from individual residential decisions can be quite inefficient both from the point of view of aggregate growth and in the Pareto sense, especially in the long run." Finally, he notes that "because of the cumulative nature of the stratification process, it is likely to be much harder to reverse once it has run its course than to arrest it an early stage."

Within the framework of the regional economic commons, we believe that sharp and pervasive disparities create a drag on and impair the productivity and performance of the regional economy. In a complex, highly interdependent economic system, therefore, widening disparities do matter to all parts of the regional economic commons.

There is a critical need for further research that focuses on the effects of sharp disparities within regional economic commons on the economic performance of the region. To address this issue effectively, research must focus on the region as an economic system rather than a mosaic of jurisdictions with jurisdictions as the unit of analysis. This is a difficult task, particularly with the limitations of jurisdictionally organized data.

The view through the lens of the economic commons strongly suggests that sharp disparities will adversely affect economic productivity and performance of the individual regional economic commons. Because the U.S. common market is a highly interdependent system, what affects one regional economy ripples out to the system as a whole. The concern about disparities in the commons is not singular to that economy but to the national economic system as a whole.

NOTES

1. Garreau, J. (1991). *Edge City: Life on the new frontier.* New York: Doubleday.

2. Dunn, E. S., Jr. (1980). *The development of U.S. urban systems.* Baltimore, MD: The Johns Hopkins University Press.

3. Cisneros, H. G. (Ed.). (1993). *Interwoven destinies: Cities and the nation.* New York: Norton.

4. Rusk, D. (1993). *Cities without suburbs.* Washington, DC: Woodrow Wilson Center Press.

5. Weak support for the elasticity hypothesis is found in Blair, J. P., Staley, S. R., & Zhang, Z. (1996, Summer). The central city elasticity hypothesis: A critical appraisal of Rusk's theory of urban development. *Journal of the American Planning Association, 62*(3), 346-353. On the basis of their statistical analysis the authors argue that the economies of states have a stronger influence on changes in poverty rates and income than central city elasticity (p. 351). Implicit in this argument is the thesis that the state is an economy or meaningful economic unit rather than a political jurisdiction astride a set of regional economies. We believe that measurements of economic performance of states are an

outcome of the performance of regional economies, not of a separate state economy. The authors' findings do suggest, however, the importance of understanding the functional relationships and linkages among local economic regions and their effects on economic performance of linked regional economies.

6. Swanstrom, T. (1996, May). Ideas matter: Reflections on the new regionalism. *Cityscape*, 2(2). Swanstrom argues that the greatest weakness of the new regionalism debate is that it is "trapped in the terms of the liberal/conservative debate between more government or freer markets" (p. 15) and "between a new layer of regional governments or governmental fragmentation" (p. 13). We believe that this characterization and generalization is not correct. Regional government or even more government is not the solution that fall naturally from the recognition that the regional economy is a collective good or commons. Rather, it is the growing awareness of the need for governance, not government restructuring: collaboration among new jurisdictions, not metropolitan government. The issue is not liberal or conservative, not more or less government, but rather how existing governments can work together to tend the economic commons.

7. *1994 Economic Report of the President* (1994). Washington, DC: Government Printing Office. (p. 26).

8. See, for example, Savitch, H. V., Sanders, D., & Collins, D. (1992). The regional city and public partnerships. In R. Berkman et al. (Eds.), *In the national interest: The 1990 urban summit* (pp. 65-77). New York: Twentieth Century Fund Press. See also studies discussed in Chapter 3.

9. Benabou, R. (1994). *Education, income distribution and growth: The local connection* (Working Paper No. 4798). Cambridge, MA: National Bureau of Economic Research.

10. Benabou, R. (1994). *Education, income distribution and growth: The local connection* (Working Paper No. 4798, p. 3). Cambridge, MA: National Bureau of Economic Research.

11. Benabou, R. (1994). *Education, income distribution and growth: The local connection* (Working Paper No. 4798, pp. 28-29). Cambridge, MA: National Bureau of Economic Research.

THE GLOBAL COMMONS

It is not just on a national scale that the nationalist economic paradigm focuses economic policy. Nationalist economics is also the lens that focuses international economic policy in the United States as well as in most other nations. The very narrowest version of this lens reveals the international economy as an assortment of autonomous nations competing in the global marketplace. Under this paradigm, nations act in concert only to set the "rules of the game" through devices such as the General Agreement on Tariffs and Trade and to maintain reasonable exchange-rate stability through international institutions such as the International Monetary Fund.

Clearly, many nations, institutions, and people have moved beyond this narrow version of the nationalist paradigm—witness the increasing number of multinational arrangements to reduce trade barriers and, in more advanced cases, act in concert on matters of economics. These include free trade areas designed to reduce trade restrictions—for example, North American Free Trade Agreement (NAFTA); customs unions that seek to create common external trade policies while reducing barriers; common markets that encourage the free movement of factors of production across national boundaries within the market area; and economic unions, the most advanced form of international cooperation, designed to achieve economic integration of national economies by creating common economic policies (e.g., the European common market).[1]

These forms of international economic cooperation reflect the growing fatigue of the nationalist paradigm. We can interpret them as modified forms of this prevailing paradigm, reflecting the adaptation of the paradigm to the recognition of an emerging economic reality.

In this chapter, we look at the global economy through the lens of the regional paradigm. It is a strikingly different view of economic reality than the one that currently guides the policy and practice of national governments as they address the global economy.

As we have stated, the stakes in selecting the proper perspective for economic policy are extraordinarily high. The costs of being wrong are staggering. Just as a narrow nationalist view undermines national economic policy making, it also distorts our view of the international economy and misleads us in policy and action on a global scale. What is needed is a meaningful paradigm of the global economy that moves beyond this limited and limiting view of independent nation-states seeking national economic advantage.

NATIONALIST PERSPECTIVE: GLOBAL AUTONOMY

The nationalist perspective of the global economy is founded on the assumption of congruence among national boundaries, national governments, and national economies. Under this critical (and unsustainable) assumption, all boundaries are aligned: the geography of the nation, the real economy, the regulatory and monetary economies, the national political jurisdiction, and, hence, the political economy of politics and economics (Figure 8.1).

This symmetry, with its orderliness and simplicity, partially explains the inherent appeal of the paradigm and the inherent difficulty in stepping outside it to achieve a more realistic perspective. Chapter 6 argued that the nationalist paradigm and its core assumption of congruence mislead policymakers and result in damage to the economies

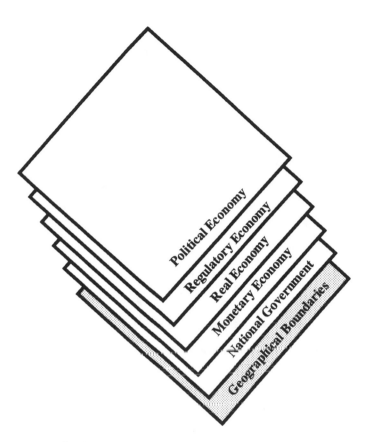

Figure 8.1. Nationalist Congruence

of regions, the national system, and, hence, national common markets. The damage of the nationalist paradigm in domestic policy is compounded by the distorted view it provides of the global economy and the resulting misdirection of international economic policy.

A highly conceptualized visualization of the global economy through the lens of the nationalist paradigm is presented in Figure 8.2. This visual construct is based on the assumption of a flat, homogeneous plain—that is, a uniform terrain uninterrupted by intervening oceans, protruding mountain ranges, inconvenient rivers, and other vexing natural barriers that create differences across space.

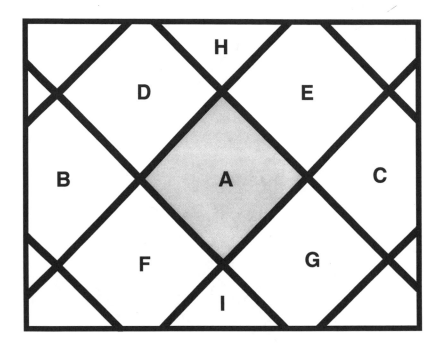

Figure 8.2. The Nationalist Perspective: The Global Political Economy

This stylized representation of the global economy communicates the tidiness and order of the global worldview through the nationalist paradigm. Each nation (represented by diamonds in Figure 8.2) has discrete, unambiguous, and congruent borders of geography, jurisdiction, and economy (per Figure 8.1). Each nation confronts an international venue composed of like nations with like congruent boundaries. Consequently, the nationalist paradigm provides for an orderly, symmetrical, and, conceptually at least, simple vision of political and economic reality.

In this world of congruency, nations and their national economies confront other nations in relatively unambiguous cross-border trading relationships. Terms of trade between autonomous national economies are determined by the comparative cost advantages to each nation. Exchange rates, the ratio of the relative value of two countries' currencies, reflect the bilateral terms of trade between the two. Over time, nations are forced to achieve a balance of trade by the inexorable discipline of the market.²

The pervasive assumption under the nationalist paradigm is *economic autonomy:* nation-states and their discrete, autonomous national economies competing in global economic warfare. This assumption of economic autonomy permeates the paradigm both in its line of sight outward to the global economy and, turning inward, in its extensions to state and local governments. The nationalist (or more broadly, the statist) paradigm assumes that jurisdictions are distinct economies in economic competition with all other jurisdictions; states and local governments are discrete and real economies in economic competition with all other states, both within the United States and in the international economy. The simplistic symmetry of the nationalist paradigm permits this pivotal assumption of autonomy of governments and economies in nationalist economic thought and public policy.

REGIONAL PERSPECTIVE: GLOBAL INTERDEPENDENCE

The lens of the regional paradigm provides a very different view of the global economy and a dramatically refocused policy perspective. This view of the economic reality of the international economy is significantly more complex, less symmetrical, and, therefore, more difficult to organize through theories than the dominant nationalist economic paradigm. It is in the global economic arena, however, that the fatigue of the nationalist paradigm is most apparent.

Figure 8.2 focused on the political economy of nationalism in the global economy. To achieve a similar perspective through the lens of the regional paradigm requires a series of overlays (Figures 8.3-8.10). Each of these overlays represents an important step in achieving an understanding of the regional global perspective and, more important, the implications of this perspective for thinking about economic policy. In this process, we again initially undertake a radical separation of economics and politics before re-fusing them in Figures 8.9 and 8.10. This process of separation of polities and economies is essential in stepping beyond the blinders of the nationalist paradigm.

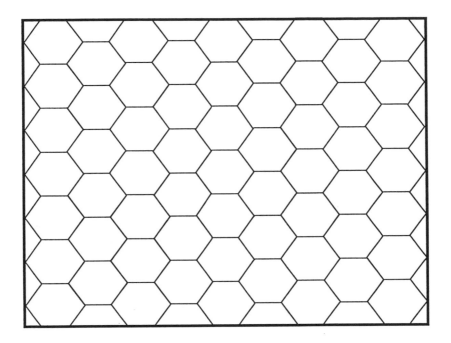

Figure 8.3. Global System of Local Economic Regions

THE GLOBAL ECONOMY

We start with the assumption of a flat homogeneous plain undifferentiated by natural geographical divisions. Local economic regions, the basic economic building blocks of the global economy, are represented in Figure 8.3 by hexagons. For simplicity's sake, these local economies completely exhaust or cover the geographical space of the global system.

Local economies are the core components of a complex, interdependent global network of economic systems. Economic systems, therefore, are complex, interdependent networks of local economic regions. One economic system is delineated from others by the patterns, intensity, depth, and complexity of the linkages and interdependence among the local economic regions that comprise the system. In other words, economic systems are systems of economically integrated local economies.

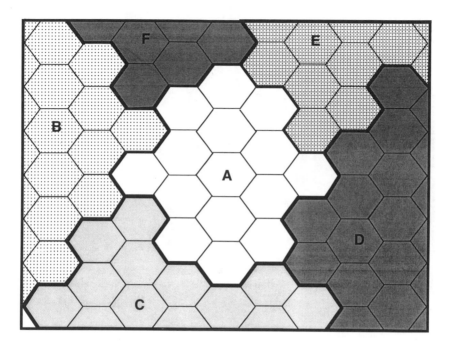

Figure 8.4. Regional Perspective: Global Network of Economic Systems

In Figure 8.4, each of the shaded areas represents an economic system composed of a network of local economies (Systems A-F). This figure, therefore, depicts a global economy composed of a network of economic systems that spans the globe.

Global economic systems are not the discrete, autonomous, independent economies of the nationalist paradigm. Boundaries of economic systems are not clearly delineated nor sharply defined as in the nationalist model. The sharply defined boundaries depicted in Figure 8.4 do not exist in reality. Rather, economic boundaries are blurred, with economic systems and, indeed, local economies flowing into and merging with one another in ways that cannot be captured in cartographic images. There is no clear line of demarcation between or among economic systems at which trade barriers can be enforced or at which the flow of factors of production and goods and services can be constrained. Rather, trade barriers are creatures of politics and political jurisdictions

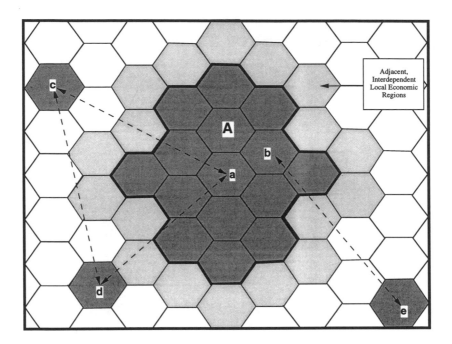

Figure 8.5. Regional Perspective: Global Economic Interdependence

enforced at points of embarkation and debarkation, as well as within the nation, or through threats of sanctions or retaliation.

Local economies and their economic systems have a high degree of interdependence with contiguous economic regions resulting, in part, from shared boundaries and proximity. In Figure 8.5, for example, the majority of local economic regions in Economic System A are adjacent to, and have shared boundaries with, the local economic regions of other economic systems (illustrated by darker shaded regions adjacent to boundaries of the highlighted economic system).

The image of the global economy as a complex network of economic systems arrayed on a geographical plane—with intensity of interdependence and efficiency of linkages resulting from proximity—is helpful but inadequate. Economic reality is significantly more complex. Intensity of interdependence among national systems is not an outcome of proximity or propinquity but of markets, materials, cost advantages, technologies, global politics, global conflict, history, and so on. Two or more local economic regions in two or more noncontiguous economic systems can be linked through economic relationships as deep and intense as those of adjacent local economies in adjacent economic systems. This is illustrated in Figure 8.5 by the dashed arrows connecting the distant (shaded) local economic region (e) to the local economic region (b) of Economic System A.

The following is an essential point: Economic relationships and economic interdependence among local economies in different economic systems transcend proximity in the globalizing economy of the latter part of the twentieth century.

Also, interdependence is not constrained to the bilateral relationships between local economies. Relationships among local economic regions in different economic systems can be trilateral (Figure 8.5; Regions a, c, and d) and multilateral (linkages among multiple economic regions in contiguous and noncontiguous economic systems). These complex patterns of noncontiguous interdependence are an essential facet of the current and future global economy.

BEYOND INTERDEPENDENCE: ECONOMIC INTEGRATION

The term *economic integration* has a dual meaning in the literature of economics. The first is founded on the interdependence of industrial sectors in an economy and the vertical and horizontal integration of industries. The second meaning is "the joining together of economic activities, especially the trade of several countries"—for example, through free trade areas, customs unions, common markets, and federations of national economies.[3] The first meaning is economic. The second is a concept of "political economy." In our discussion, we are using the term primarily in the economic sense.

Integration of economies and economic systems occurs when one or more local economic regions become an integral part of two different economic systems. In the absence of jurisdictionally imposed boundaries, adjacent economic systems may integrate or merge, with some local economies being parts of each and held in common.

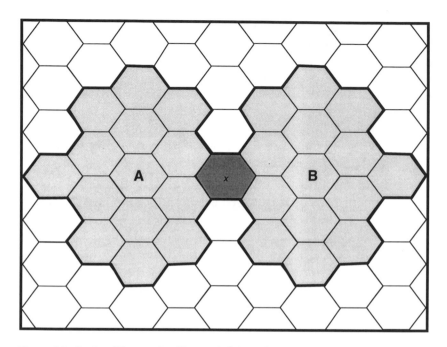

Figure 8.6. Regional Perspective: Economic Integration

In Figure 8.6, Economic Region x is a building block of two economic systems. These two systems are economically integrated. Although the figure depicts only one economic region held in common, the overlapping of two (or more) economic systems may well include multiple local economies (Figure 8.7). Indeed, as the logic and the impetus of economic integration increase, economic systems may merge with growing economic linkages that overwhelm the patterns of interdependence that make the systems distinct.

Proximity and contiguity are also not necessary for economic integration. In Figure 8.7, the dashed lines depict economic integration among three noncontiguous economic regions (x, y, and z) in different economic systems of the global economy.

These patterns of interdependence and integration of economic systems and local economies provide a perspective of the global economy that stands in stark contrast to the nationalist paradigm.

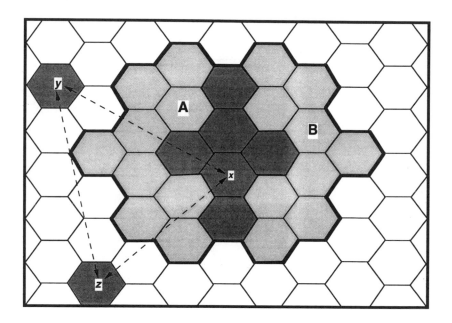

Figure 8.7. Regional Perspective· Integration of Economic Systems

From this fresh perspective, the international economy is the global network of interdependent and economically integrated economic systems and their component local economic regions.

THE GLOBAL POLITICAL ECONOMY

In the preceding sections, we have taken the final step in building the case for the economic federalism of the regional paradigm by applying it to the global economy. It is now important to re-fuse politics and economics and to ask how this political economic perspective changes the way we must think about economics, politics, and policy making in the international arena. This re-fusing of politics and economics according to the regional perspective provides a new framework for under-

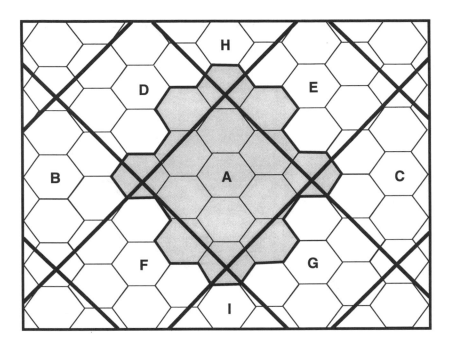

Figure 8.8. Regional Perspective: Global Political Economy (A-I Denote Nation-States [Diamond-Shaped Figures])

standing the global economy and for thinking about policy and policy making.

The risks of addressing the global economy through a restrictive lens of nationalism become readily apparent when boundaries of nation-states are superimposed on the global network of economic systems and their local economic regions. This is done in Figure 8.8, which combines or overlays Figures 8.2, 8.3, and 8.4. Figure 8.8 provides a stark illustration of the problem of noncongruence in the global economy.

For the purpose of clarity, it is useful to restate the basic theses of the regional paradigm in the context of Figure 8.8. The real national economy is the economic system of local economies. These economic systems spill across the jurisdictional boundaries of nation-states (represented as heavily lined diamond shapes in Figure 8.8). The framework of nationalist economics views only that part of the economic system contained within national boundaries as the national economy. The regional paradigm, in contrast, recognizes the part of the economic system contained within national boundaries as the national common market but not as the economic system as a whole.

Where economic systems spill across national political jurisdictions, national economies cannot be the independent economic entities of the nationalist economic paradigm. Through the lens of the regional paradigm, we see a very different economic reality. A local economy that is partly within the political jurisdiction of Country A, for example, is also partly within the jurisdictions of adjacent nations. Where this occurs, both nations have a vital economic stake in the local economy, even if that economy is more tightly tied to the economic system of Country A. These two nations are thus "joined" in an economic relationship of mutuality and interdependence rather than the singular and unrelenting competition of nationalist economics and politics.

The level of complexity in the juxtapositions of nations, common markets, and economic systems can be significantly greater. The economy of a nation, its common market, might be composed of parts of two or more economic systems (Figures 8.9a and 8.9b). In Figure 8.9a, for example, the national common market of Nation A includes parts of two economic systems.

In Figure 8.9b, Nation A stands astride parts of four economic systems. In this figure, no country incorporates the major part of any economic system; common markets are composites of parts of several systems. The meaning of a national common market, in this case, takes on a very different cast. Common markets are highly interdependent with those of surrounding nations and are highly dependent on the performance of economic systems that cross national boundaries. They are not the autonomous national economies of the nationalist paradigm.

As a result, where two or more systems are economically integrated, or are in the process of integrating, the risks of economic nationalism—a perspective that is unaware of or ignores this intertwining of economies —are great (Figure 8.10). The goal of economic policy viewed through the lens of the regional paradigm, however, will be to promote economic integration in the global economy. This goal will be very difficult to perceive through the lens of the nationalist paradigm, especially when these economic ties are among noncontiguous regions (x, y, and z in

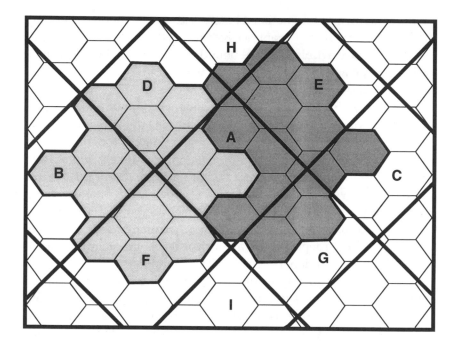

Figure 8.9a. Regional Perspective: Interdependence in the Global Economy (A-I Denote Nation-States [Diamond-Shaped Figures])

Figure 8.10). The resulting policy likely will impede rather than facilitate economic integration.

Where economically integrating or integrated economic systems transcend the boundaries of two or more nations, therefore, each nation has a critical stake in the performance of these systems and will be significantly affected by the economic policy decisions, both domestic and international, of all the nations involved.

A narrow nationalist policy orientation, based on the assumption of economic autonomy, has great potential to impair, rather than enhance, the performance of the affected economic systems and, therefore, the national common markets of each nation. National economic policies implemented on the assumption of economic autonomy can potentially harm the common market of the implementing nation. This outcome will be counterintuitive through the lens of the nationalist paradigm.

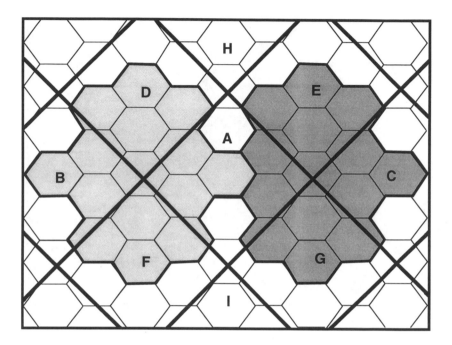

Figure 8.9b.

THE GLOBAL COMMONS

Viewed through the regional paradigm, the global economy is the
network of interdependent economic systems and their component
local economic regions, and the global polity is the lattice of nation-
states. The global political economy, therefore, is the overlay of nation-
states and global economic systems.

This global political economy has two key characteristics. First,
polities and economies are not congruent in the global political economy. The clear, crisp jurisdictional boundaries of nation-states are not
aligned with the blurred and less distinguishable demarcations of
global economic systems.

Second, the global political economy is highly interdependent. This
global interdependence takes two forms. One is more purely economic

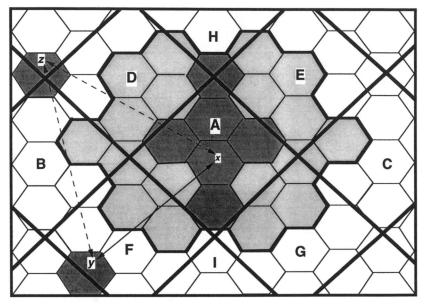

Diamond shaped figures (letters A-I) denote nation states

Figure 8.10. Dimension of Economic Integration (A-I Denote Nation-States [Diamond-Shaped Figures])

in nature than the second and is based on the high degree of interdependence and integration of economic systems in the global economy. The other results from the very nature of the global political economy. Nations are economically interdependent because they share or hold in common parts of economic systems within their common markets. All nations incorporating part of a given economic system within their boundaries, therefore, have a mutual stake in the performance of that economic system. The noncongruence of national common markets and global economic systems creates global economic interdependence.

Economic mutuality and interdependence are the central characteristics of the regional paradigm, an assumption in stark contrast to the core premise of economic autonomy of the nationalist paradigm.

The concept of a global economy composed of multiple economic systems and many nation-state stakeholders again poses the dilemma of the economic commons. The question is how to nurture the global economic commons in a world dominated by narrowly perceived nationalist economic interests.

The regional paradigm focuses the critical issue of the fragmentation of economic systems by national boundaries. The risks of the national paradigm's assumption of economic autonomy become obvious. Policies focused singularly on the perceived economy within national boundaries, and designed to promote the competitive advantage of one nation, inevitably will have adverse effects on the performance of each of the economic systems that fall partially within the geographical boundaries of the national jurisdiction.

When boundaries of nations do not incorporate major portions of any economic system, and economic systems are overlaid by the boundaries of two or more jurisdictions, the danger in viewing the global economy through the lens of the nationalist paradigm and making economic policy in this context is very great indeed. Nations that allow their national boundaries to define both their economy and their "national interests"—and then proceed to make policy according to this limited and limiting view of the world—can harm the global commons.

Equally important, they will implement extremely shortsighted domestic economic policies that impair the local economic regions of their common markets. We have already pointed to the problem of making economic policy based on national averages of economic indicators. The complexity and risks of economic policy making triggered by national averages are greatly compounded when these averages reflect data from local economic regions in two or more economic systems falling partially within national boundaries.

The interdependence of the global commons creates a prima facie case for nations to join together in the creation of multinational common markets that can, to a greater extent, encompass one or more economic systems. This is analogous to the issue of multiple local governments overlaying the regional economic commons. How do governments cooperate to nurture the commons of economic systems?

This global political economy framework also provides a useful perspective on political realignments such as the dramatic changes in national political jurisdictions that have occurred and are occurring in Eastern Europe.[4] To the extent that the underlying causes of the collapse of the Soviet Union were economic, one cause may have been a political scale in which the Soviet common market encompassed several global economic systems. The polity thus could not sustain the internal integrity of the perceived national economy. This contrasts with the European common market, which is an effort to ratchet the political scale upward to the greater economic rationality that comes with the alignment of political and economic boundaries.

The breakup of the Soviet Union is a ratcheting downward of political scale. It appears, however, to be primarily ratcheting downward to greater political rationality defined by ethnicity, religion, and historical national identities. It does not appear to reflect a ratcheting to greater economic rationality and therefore may contain the seeds of current and future economic woes for residents of newly forming nations.

ECONOMIC ROLE OF NATION-STATES

The ideas of Kenichi Ohmae, in *The Borderless World* and *The End of the Nation State*,[5] and Jane Jacobs provide a useful counterpoint to economic nationalism and the nationalist paradigm. Ohmae and Jacobs both make an important and useful jump from a focus on the nation to a focus on local economies. Their positions are the antithesis to nationalism. In this, however, they divorce politics and economics in an unrealistic and potentially dangerous way. Furthermore, Ohmae and Jacobs fail to take the second essential leap, which is to see both the global economy and the local economic regions as "commons."

The regional paradigm provides a quite different view of the economic reality of the global economy from both the nationalist paradigm and the "economism" of Ohmae and Jacobs. It provides a framework for understanding the global political economy that has important implications for policy and policy making.

National governments will continue to have a critical role in the global economy. Ignoring the role of nations in new paradigms of the international economy is shortsighted and unrealistic. Nation-states, although no longer the dominant economic units of the international economy under the regional paradigm, continue to be an important and influential subsystem of the emerging global political economy.

Nations are political, not economic, entities and have political functions separate from and transcending their economic roles. Furthermore, national sovereignty provides the power to facilitate or truncate cross-economic systems. The power of nations to regulate trade, set exchange rates, establish tariffs, regulate industry, and administer national industries ensures that nation-states will continue to be powerful factors in the emerging international economic order.

The role of international organizations also appears quite differently through the lens of the two paradigms. Under the nationalist paradigm, the focus of multinational organizations is primarily to establish and enforce the "rules of the game" in international trade. Through the lens of the regional paradigm, the fates of economic systems and national economic commons are tied together. International institutions, therefore, will pursue the goal of enhancing the global commons, building on interdependence and economic linkages.

THE ECONOMY OF PRODUCTION IN THE GLOBAL COMMONS

We have described a global political economy of three interacting systems: the regional economic commons, the common markets of nation-states, and the global commons. The spatial organization of the production economy, the economy of goods and services, is arrayed within and across these interdependent political economies of the global commons. This relationship, unfortunately, is not clear.

An important debate is occurring on the emerging organization of production in the global economies as well as the effects on nations and regional economies. One position argues that we are now crossing a "second industrial divide" that is radically changing the nature of economies and regions or districts.[6] Central to this position is the thesis that globalization and its attendant complexities increasingly make it difficult for large, unwieldy corporate organizations to compete effectively in the emerging global economy. New technologies support the development of smaller enterprise. Accordingly, in the new industrial milieu, smaller enterprises, networked within "districts," are the emerging form of production organization. The networking of these enterprises within regions or districts permits capture of relevant economies of scale, allowing these firms to be competitive in national and global markets. It is also argued that increasingly distinctive consumer tastes and preferences across regions make it difficult for large, highly concentrated corporations to address these markets, thus creating opportunities for smaller firms to capture these specialized "niches."

The implications of this position are appealing for those who care about regions and the efficacy of local economic development strategies. Economic growth is rooted in smaller, innovative enterprises, and these are organized in district networks. With an understanding of these clusters in any given economic region, effective economic development strategies can be designed and implemented to promote these enterprises. This orientation also minimizes the sense of linkages with other regions by lending support to the view that regions or districts can be autonomous and competitive.

A powerful counter to this thesis is posed by Bennett Harrison in his book, *Lean and Mean: The Changing Landscape of Corporate Power in the Age of Flexibility.*[7] Harrison suggests a quite different reality. After a period of expansion and diversification, he asserts, corporate strategies now dictate scaling back to core competencies to become "lean and mean" while building sophisticated production networks or global webs. Concentrated economic power, according to Harrison, is not diminishing but rather is changing its shape through strategic alliances, technology agreements, and cross-licensing agreements among firms as well as government agencies, financial institutions, universities, and so on. Smaller firms serve primarily as suppliers and subcontractors in these emerging global production networks rather than as primary drivers of economic growth. Harrison wrote,

> The emerging global economy remains dominated by concentrated, powerful business enterprises. Indeed, the more the economy is globalized, the more it is accessible only to companies with global reach. The spread of networking behavior signifies that the methods for managing that reach have changed dramatically, not that there has been a reemergence of localism as others have charged.[8]

It is early in this debate, as well as in the "second industrial divide" or the "third shift" in capitalist development[9] (a term used to characterize the rise of densely networked production economies), to understand the emerging spatial organization of economic production. In both, production is grounded in economic regions and their national common markets. It is clear, however, that one implies greater stability for regions, whereas the growth of major economic networks may entail less attachment to place for major corporations, smaller firm dependency as suppliers and subcontractors, and, hence, greater economic volatility for economic regions.

In both cases, however, the national and international grids of regional economies and their larger economic systems will be a primary determinant of the spatial organization of economic production as firms, great or small, seek access to markets, suppliers, and high-productivity environments offering high-quality workforces and requisite infrastructure.

THE REGIONAL ECONOMIC PARADIGM

Effective policy requires that government be able to see the international economy clearly. Under the nationalist economic paradigm, the line of sight again is seemingly unobstructed, from the national government to other national governments. This is perceived as a useful view of economies as well as governments. The international economy, according to this paradigm, is the global configuration of national government economies. This sense of clarity of vision, however, is an illusion.

Through the lens of the regional paradigm, the global economy appears very different. In Chapters 5 through 7, we have systematically presented the elements of the paradigm, beginning with (a) the basic building block of the local economic region and its political economy, the regional economic commons; building to (b) the economic system and the political economy of the national common market; and, finally, (c) the global economy and the political economy of the global commons.

From the viewpoint of a local economic commons, one looks outward to the national common market, and through that, to the global economic commons (Figure 8.11). The line of sight does not end there, however. What is seen through the lens of the regional paradigm is another regional economic commons in another common market.

Finally, we see the essence of the regional economic paradigm. There are three interdependent economic systems—local, national, and

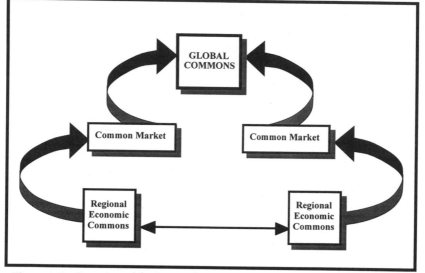

Figure 8.11. Overview of the Regional Economic Paradigm

global—and three interdependent political economies. No one system can clearly be perceived and understood without understanding these three tiers and the interactions among them. The basic economic relationship, however, filtered through the national and global tiers, is between and among local economic regions and their regional economic commons.

The key and critical linkage in the regional paradigm is from one local economy and local economic commons to another, whether they are in the same or different common markets or economic systems or both. The clouding veils of national jurisdictions, as seen through the nationalist paradigm, obscure this line of sight between local economic commons. The regional paradigm pierces these veils and clearly reflects that the critical economic relationships are between or among regional economic commons.

The "call to arms" of local economic development is to "think globally and act locally." Understanding the regional paradigm changes gives new content and meaning to this aphorism. Globally competitive regions will think and act both locally and globally.

This understanding also provides clear directional signals for governments, from local to national: Act to build, nurture, and sustain linkages among regional economic commons, domestic and international; avoid actions that truncate or diminish this linkages; and beware the simple and often misleading answers of statism and the nationalist paradigm at all levels of government.

NOTES

1. This discussion of forms of international economic cooperation and the list of trading blocs that follows are abstracted from Dicken, P. (1986). *Global shift: Industrial change in a turbulent world* (pp. 146-147). New York: Harper & Row. In the postwar period, a variety of trading blocks have arisen to enable coordinated policies and concerted actions. Among these are the European Economic Community, European Free Trade Area, Latin America Free Trade Area, Central American Common Market, Andean Common Market and Community, Caribbean Economic Community, Union Douanière et Economique de l'Afrique Centrale, East African Community, Communauté Economique de l'Afrique de l'Ouest, Economic Community of West African States, and the Association of South East Asian Nations. To these can be added the nations included in the NAFTA.

2. Under the logic of this system, a nation cannot have a long-term deficit or surplus in the balance of trade with another country. Over time, deficits or surpluses will affect the exchange rate in a manner that affects the costs of exports or imports valued in terms of the currencies of the other nation. For example, the United States, other things being equal, will be unable to run a long-term deficit in its balance of trade with another country because the increased supply of dollars in the other country will decrease the value of the dollar relative to the currency of that country—that is, the exchange rate will shift in favor of the nation with the trade surplus. As a result, U.S. commodities will become cheaper when purchased with the other nation's currency and that nation's goods will become more expensive in terms of U.S. dollars. As the demand for U.S. goods increases, demand for the other country's products decline, thus reducing the U.S. trade deficit as well as the surplus balance of trade of the other nation. In the jargon of the economist, equilibrium will be restored.

3. See Donald, R. (1992). *Dictionary of economics.* New York: Routledge.

4. This issue was suggested by Edward Hill, Senior Research Associate, The Urban Center, Levin College of Urban Affairs, Cleveland State University.

5. Ohmae, K. (1991). *The borderless world* (p. 18). New York: Harper Perennial; Ohmae, K. (1995). *The end of the nation state: The rise of regional economics.* New York: Free Press.

6. Piore, M. J., & Sabel, C. (1984). *The second industrial divide: Possibilities for prosperity.* New York: Basic Books.

7. Harrison, B. (1994). *Lean and mean: The changing landscape of corporate power in the age of flexibility.* New York: Basic Books.

8. Harrison, B. (1994). *Lean and mean: The changing landscape of corporate power in the age of flexibility* (p. 12). New York: Basic Books.

9. Gereffi, G., & Hamilton, G. G. (1990, August). *Modes of incorporation in an industrial world: The social economy of global capitalism.* Paper presented at the Annual Meeting of the American Sociological Association, Washington, DC Cited in Harrison, B. (1994). *Lean and mean: The changing landscape of corporate power in the age of flexibility* (p. 219). New York: Basic Books.

CHAPTER **9**

INTERGOVERNMENTAL ROLES IN ECONOMIC POLICY MAKING

Two noncongruent federalisms—one political, the other economic—provide a difficult framework for making good policy. The greater the mismatch, the less facilitative is the framework and the more difficult it is to establish effective governance and good policy. One appeal of the nationalist paradigm is that its simple equating of state and economy avoids this sense of mismatch.

The lens through which the economy and the political economy are viewed fundamentally shapes economic policy and governance. After all, it is the reality that is defined through this lens that influences the allocation of policy roles, the search for appropriate governance mechanisms, and the choice of solutions. A good lens leads to appropriate questions and the possibility of useful answers. A bad lens leads to inappropriate questions and, almost certainly, to answers that are not useful and often counterproductive.

A great threat to good policy and governance is a refusal to recognize that the political and economic worlds are not spatially aligned, a refusal to accept complexity and to give up the nationalist lens. A government that thinks its boundaries circumscribe a functioning economy and acts accordingly is in a position to do much damage.

Equally dangerous is the belief that governments can do nothing well nor accomplish anything good. Discussion about the importance of local economies in the global economy, for example, sometimes focuses on the increasing need for regional action and the decreasing importance of the nation-state. Kenichi Ohmae, in foreseeing and urging *The End of the Nation State*, says that his major concern is that local entrepreneurs in the public and private sectors should be able to connect with the global economy.

Like Jane Jacobs (and also like a currently fashionable political perspective in Washington), Ohmae seems to think the nation-state can only get in the way of this crucial economic activity. The central government's role, according to Ohmae, should be to "facilitate" the activities of "region states." Nevertheless, Ohmae does note there are "kinds of value that only a central government can appropriately provide— military security, for instance, a sound currency, infrastructure standards, and the like."[1]

As noted previously, Ohmae's arguments and others like it are an exercise in simple economism, based primarily on the views of the chief executive officers of transnational corporations and eager entrepreneurs. According to this line of thought, government is expendable or, at best, only useful when it is at the service of an exciting global economic dynamic.

Our argument here is quite different. The common market regionalist view of the political economy calls for important innovation at all levels of government (and, indeed, internationally) and among all important actors in the political economy. Moreover, there is no structural fix—no abolishing of one type or level of government or establishing of some new one—that can accomplish what is needed. The issue is not how to minimize government but how to reconstruct it.[2]

For example, economic policy roles formerly held exclusively by the federal government may need to be shared, but effective federal performance of the appropriate aspects of those roles will be necessary to the successful functioning of local economies and regional economic commons (RECs). (This may also be Ohmae's view. His exuberant advocacy of the economic dynamic may exaggerate the antigovernment theme. We hope this is so. His strong objection, however, is to "protectionism" emanating from Washington. It is unclear why this will not occur at the city or region level.)

Thus, the economic paradigm and the new political economy described in the previous chapters embody the following three fundamental challenges for policy and governance:

1. How will governance and policy be accomplished for each regional economic commons?
2. How will governance and policy be accomplished for the common market?
3. How will governance and policy be accomplished for the global commons?

The challenges are similar in all three cases: in a complex system, in which political and economic aspects are not congruent, how can an effective political economy be created within which to conduct governance and make good policy? At each level, there is considerable interdependence among the units, both political and economic. Furthermore, the three "levels" are not autonomous; they interact. In effect, all three are embodied at each level; there are multiple as well as unitary interests in each commons. In other words, each question is a window onto the whole system.

Although this makes analysis complicated, it also means that the whole system may be affected by clear and consistent action at any point within it. This duality—that is, the need to recognize complexity and the opportunities for specific, important action—pervades this and subsequent chapters.

Chapter 8 addressed the third question, describing a new framework for the global and international economy and the challenges it contains for economic policy making. Answers to the first and second questions will be explored in Chapters 10 and 11. To get to those, however, we first need to address, in this chapter, the salient question in contemporary American federalism: How should roles and relationships along the federal-state-local axis—the normal frame of three-tiered vertical federalism—be allocated?

Establishing effective governance at one level in the new political economy will significantly affect our ability to do so at the other levels. In the process, the relationships among federal, state, and local governments will be transformed. For example, a major oversight in many discussions of "metropolitan governance" is their exclusion of the crucial roles that state and federal governments must play. The "metropolitan question" is treated, in such discussions, as if it were a separable topic, a distinct function. The assumption is that the appropriate role for upper levels of government is to provide a set of carrots and sticks to induce appropriate local behavior. This is not enough, however, nor is it the right way to approach the issue. Regional governance and regional economic policy making pervade the spheres of federal and state (as well as local) responsibilities. All levels of government have much at stake in the outcome.

Vertical federalism has received much recent attention, especially in the rather misshapen form of the "sorting out" debate. Elected officials and scholars have devoted considerable energy and ingenuity to proposing schemes to separate the various functions of the public sector and then to divvy them up among federal, state, and local governments.

Federalism and even "intergovernmental relations," however, are richer concepts than the recent sorting out discussion allows. As the previous chapters made clear, "national" does not necessarily mean only "federal government" in economic policy making. Indeed, even from a purely political perspective, national does not mean "central." In the political economy, national may also mean the system of RECs linked in the common market. In this latter context, the standard vertical view of three levels of government—federal, state, and local—must be flattened into a horizontal set of relationships. This is new terrain for most contemporary governmentalists, scholars and politicians alike. The common market framework heightens the importance of the set of intergovernmental relationships that are necessary to construct the political economy for the regional economic commons.

Stated another way, issues of intergovernmental relations most often are framed from the top down. This is logical and apparently adequate within the nationalist paradigm of political economy. Shifting to the new paradigm proposed here, however, makes clear that the top-down approach does only half the job; the other half is the perspective from the bottom up. If the great machine of the political economy is to work, the piston of federalism must make both strokes.

This chapter, therefore, argues (a) that the choice of political economy paradigm matters hugely in answering the federalism question, (b) that sorting out roles will not help us address the economic problems that confront us, and (c) that effectiveness and democratic values rather than obsessive neatness are the preferable basis for building a functioning federalism that can provide policy and governance for the immense economic policy challenges we face.

PARADIGMS MATTER

The answer to the question of how governmental roles should be assigned for economic policy depends on which paradigm of the economy is applied. This point is vividly illustrated by comparing the views of two thoughtful analysts, Jane Jacobs and Paul Peterson. Using radically different views of the economy, they present radically different ideas about the proper roles of federal, state, and local governments. Using a statist view of the economy, Peterson sees few economic policy roles for local governments; their job is to feather the nest of their tax base. Using a radically regional view of economics, however, Jacobs sees few roles for the federal or state governments.

Although Jacobs and Peterson have clear disagreements about governmental roles, those disagreements cannot be resolved or even understood until we grasp the economic paradigms that lie beneath them.

In his influential *City Limits*, Peterson shaped the issue distinctly by asserting that the "place of the city within the larger political economy of the nation fundamentally affects the choices that cities make." He noted that there are "crucial kinds of public policies that local governments simply cannot execute" (e.g., make war, control passports, issue currency, and impose tariffs). Lacking such powers, "cities have limits," Peterson observed, and these can be understood "by looking at the place of the city in the larger socio-economic and political context."[3]

Peterson finds "the primary interest of cities," in this framework, to be "the maintenance and enhancement of their economic productivity."[4] He then argues that redistributive policies "are not easily implemented" by local governments and that the federal government "bears the greatest responsibility" for these policies.[5] It is in the allocational and developmental arenas that local governments "are particularly active."[6] Although politics internal to the community are not "absent,"[7] the structure of the political economy largely controls the basic options available, and this range of realistic options in turn shapes the locality's politics.

Although *City Limits* is concerned with local politics and policies, its analytic frame contains a fairly conventional conception of the whole political economy. Peterson sees the "economy" as a national entity of which any city comprises an arbitrary and very small part. Therefore, cities deliver services and seek investments to enhance their positions relative to other cities. More significant, the power of any specific place to influence the economy is arbitrary and small.

Jacobs, in *Cities and the Wealth of Nations*, provides a radically different view. Like Peterson's, Jacobs's view of intergovernmental roles derives from a paradigm of the political economy. The authors' paradigms, however, could not be more different. Whereas Peterson defines economies to fit the bounds of government, Jacobs defines government to fit her view of the economy.

Jacobs argues that "city regions" are macroeconomies themselves and that they establish linkages and trade relations with other cities. (Ohmae's similarity to Jacobs is most obvious on this point.) According to this view, of course, it makes sense that a local government that encompasses the city region economy will have full economic policy powers.

Jacobs allocates all economic policy roles to the city region and none to the nation-state. Thus, for example, "as a rule," she says, national or international product standards harm economic development.[8] City regions, she adds, should be able to break away from their nation-states. Economically, a "chief advantage" of such a situation would be a "multiplication of currencies" whose relative valuation would provide constant and correct feedback.[9] This is a "theoretical possibility,"[10] she asserts—a "utopian proposal" whose brightness illuminates the ideas she is trying to examine.

Jacobs also offers some more prosaic recommendations for the world short of utopia. Taking as her touchstone the importance of the performance of the city region, for example, she opposes value-added taxes

Table 9.1   Allocating Policy Authority Intergovernmentally

	(1) Pure Nationalist	(2) Modified Nationalist	(3) Radical Regional	(4) Common Market Regional
Countercyclical fiscal	F	F	R	CFSRL
Monetary policy	F	F	R	CFSRL
International economic policy, trade, and tariff	F	F	R	CFSRL
Capital financial policies	F	F(S)	R	CFSRL
Labor policies	F	F(S)	R	CFSRL
Land policies	FSL	FSL	R	CFSRL
Infrastructure	FSL	FSL	R	CFSRL
"Climate" for entrepreneurship	FSL	FSL	R	CFSRL

NOTE: Abbreviations used: C, common market; F, federal; S, state; R, regional economic commons; L, local

and national mandates; she prefers performance standards. In practical terms, she seeks ways for a nation "without damage to itself as a political unit . . . [to make] a little more room in itself for open-ended economic drift," which she sees as "like biological evolution" in contrast to the rigidities of the nation.[11] (As noted in Chapter 2, this radical separation of economics from government, this economism, was Jacobs's major achievement in *Cities and the Wealth of Nations*. Her subsequent effort to reconstruct political economy was perhaps less successful.)

Using Peterson's nationalist paradigm, Jacobs's views make no sense and vice versa. Reviewers of Jacobs's book looked at it through the lens of the conventional nationalist paradigm: One reviewer said her views are "anarchism by another name—decentralization."[12] Similarly, scholars of urban politics have wrestled with the implications of Peterson's emphasis on the structural limits to local choice.[13]

Such arguments about federalism and intergovernmental relations can best be understood if the underlying paradigm of the political economy is made explicit.

To make this point in a different way and with greater specificity, Table 9.1 presents a list of economic policy topics that are allocated intergovernmentally according to different paradigms of the political economy.

Reading from left to right, Table 9.1 traces the process that Kuhn called a "paradigm shift."[14] In column 1, the "pure nationalist" paradigm dominates thought. In column 2, anomalies accumulate and are accommodated into a "modified nationalist" paradigm. By the time we get to column 3, the accommodations have so thoroughly distended the old paradigm that a new paradigm is suggested in a rather pure, breakthrough form: the "radical regional" view. These ideas are then applied and developed into a usable form, the "common market regional" paradigm described in this book (Table 9.1, column 4).

Under the pure nationalist model (Table 9.1, column 1), all topics are allocated to the federal government, with state and local governments participating in only a few categories. There is no decision making at the level of the regional economic commons (R, Table 9.1)) or the common market (C, Table 9.1). States and local governments come out poorly; relative to the federal government, they have fewer roles and no exclusive ones. This is how the nationalist economic paradigm creates a powerful position for the federal government.

Under the modified nationalist view (Table 9.1, column 2), state and local governments play a proportionately greater role. They still have no power over fiscal, monetary, or trade policies and no dominant role regarding any policy tool. They nonetheless often are held accountable by their electorates for the performance of their "economy," although no rationales exist for its delineation in any but an arbitrary way. Similarly, state and local governments have little basis for decisions about how and whether to affect their economy. Moreover, when push comes to shove—which is often the case in the economic policy realm— the Feds hold the trump of the primacy of the national economy and do not hesitate to play it against the merely "special interests" of state and local concerns. Column 2 in Table 9.1 best describes the current state of affairs in the United States (and best represents Peterson's view as well).

The pure or radical regional paradigm (Table 9.1, column 3), presents an extremely different pattern. According to this view—the "theoretical possibility" that Jacobs put forth—federal, state, and local governments as we know them have no economic roles at all assigned to them. Instead, a regional level of government monopolizes policy allocation.

In column 4 in Table 9.1, federal, state, and local governments are brought back into the picture. This column reflects the common market regionalism described in the previous chapters. Roles in formulating and implementing policy in all areas are shown as shared among all sets of policymakers. Dictating this approach are the interlinked nature of the U.S. common market and the noncongruent relationship between the two federalisms.

Like the Peterson-Jacobs comparison, this display, across the four columns in Table 9.1, shows that the political economy paradigm one uses directly shapes the answers one gives to the question of how best to allocate roles among governments.

BEYOND THE FANTASY OF SORTING OUT

In contrast to the currently dominant focus on sorting out functions among levels of government, we argue for a different view of economic policy making, one based on adaptation and shared roles. A priority, of course, is that policy making account for both the local economic region and the economic system. The question, arising from the noncongruence described in Chapter 6 and pursued further in Chapters 10 and 11, is how to focus effectively on both geoeconomic targets in the context of the United States' political federalism.

Common market regionalism and its associated view of shared intergovernmental roles run counter to currently dominant views about economic policy and federalism. Regarding policy, the nationalist economic paradigm provides a rationale mainly for federal economic policy. Despite increased state and local government activity in economic development and efforts to develop small-area economic interventions, no widely accepted economic framework now exists to guide and justify such actions. The regional economic paradigm provides such a framework.

Regarding federalism, the dominant rhetoric of the past two decades has highlighted the need to clarify the intergovernmental system, especially by separating governmental roles. In the economic policy realm, efforts to separate and alter the roles of the different levels of government have been shaped by the dominant nationalist paradigm. As with Peterson's analysis and columns 1 and 2 in Table 9.1, there is consequently no coherent rationale for assigning state or local economic roles (and certainly nothing for the common market or the RECs). Thus, the sorting out proposals are based on rationales that are either arbitrary or essentially defined and driven by management or budget goals. At worst, they are a scrim behind which the real action takes place—ending federal involvement or shifting a burden to state or local governments or both. There have been no proposals by major political actors, however, to "sort out" the key roles of the economic policy inner sanctum: fiscal, monetary, or trade policies.

These discussions have both shaped and been shaped by several presidents, all of them using the lens of the nationalist paradigm of the political economy. Conlan's *New Federalism* carefully distinguishes between Nixon's "New Federalism," which aimed at "rationalizing and decentralizing an activist government," and Reagan's, which made a cause of "rolling back the modern welfare state itself."[15] As a presidential candidate in 1988, George Bush referred to Reagan's New Federalism as an "effort to shift much of the responsibility for local needs from Washington to our states and cities." He endorsed the attempt "to return to the proper roles of state and local government,"[16] but he did not advance the agenda substantively. Presidential candidate Bill Clinton spoke in 1992 as a governor who understood and intended to put in place ways of proceeding that would respect state and local variety and their need for flexibility. His administration's concern has not been to sort out policy arenas but to provide state and local flexibility in applying federal goals and programs to specific situations.

The 104th Congress, with a Republican majority following the November 1994 elections, brought a new fervor and additional ideas to what has been called the "devolution" agenda. The Congress's meshing of budget-cutting imperatives, ideological search-and-destroy missions, and "closer to the people" federalism is more complex if only because of the multiplicity of authoritative voices.

Policy analysts have also shaped and been shaped by this agenda. Among the more systematically grounded proposals for sorting out came from Alice Rivlin, former Congressional Budget Office director and former director of the White House's Office of Management and Budget. In *Reviving the American Dream*,[17] she argued that the persistent federal deficit requires a definitive and substantial reallocation of policy and funding responsibilities—that the structural budget deficit demands sorting out. This is, in effect, a way of off-loading federal burdens, but it does not make good economic policy making any more likely. If anything, favorable outcomes are less likely because the policy-making frame has substituted the condition of the federal fisc for the condition of the economy as the focus of concern. Although fiscal and economic conditions are related, they are not the same.

Whether framed in the terms of political campaigns or systematic policy analysis, these views constitute a search for the grail of a "dual federalism," a quite different set of ideas from the "two federalisms" we have analyzed here. In the dual federalism context, federal and state-local governments are separate and layered systems.

Beyond whatever constitutional appeal dual federalism may have, it also holds out a structural fix for pressing problems. The backdrop is the nationalist paradigm of an economy marked by no significant internal differentiation. Because there is no force outside of government (e.g., a local economy) that drives policy decisions, the imperatives of budgets and management can serve as the rationale for decentralization or devolution. These views are thus an escape from complex economic reality and are not likely to provide a basis for good economic policy and governance.

Contemporary sorting out is merely hierarchical; federal functions are cascaded down the vertical system. The view from the bottom is not part of the picture. The federalism piston, again, makes only the down-stroke. Is it any wonder that the machine does not work very well?

POLICY AND GOVERNANCE IN THE U.S. COMMON MARKET OF REGIONAL ECONOMIC COMMONS

In contrast to the sorting out approach, the unallocated economic policy-making roles in column 4 of Table 9.1 reflect our view that, at this point, all those roles need to be rethought and many, or probably all, need to be shared. Even the most closely guarded roles of the federal government—monetary, fiscal, and trade policy—should be opened up to voices that speak for and about the common market and the RECs. (The system needs to work better upward as well as downward.) Compared to the apparent simplicity of sorting out roles, this may be messy—in Grodzins's imagery, it is more a marble cake than a layer cake.[18] The operational criterion, however, is not neatness; it is effectiveness in achieving good policy and democratic governance for local economies and the U.S. common market.

The existing federal system contains the flexibility, theoretically at least, to accommodate all these changes through a grand variety of methods. The notion of a paradigm shift may convey, for some, a full-scale, wide-ranging, all-at-once transformation. The process of change, however, will, we think, be necessarily (and probably beneficially) incremental. Neither a grassroots revolution nor an intellectual coup d'etat are prerequisites for specific innovations to occur. Evidence of innovations that work can lay the basis for further innovation. Constructive and rewarding actions can be taken at a variety of places at all levels of the system without waiting for a cataclysmic nationwide change of paradigm. The immense internal variety of the American governmental system provides opportunities for this process of guerrilla intellectual and policy change to occur. The process has already begun. Activity is especially apparent at the regional level, although no place has achieved a fully functioning "regional commons."

Perhaps the greatest opportunity for that flexibility to be taken advantage of fully and to good effect is for governments to internalize the view of the economy and the political economy presented here. Local economic development strategies, for example, must look not only at what is happening within a locality's political boundaries but also at the larger regional economy as well. We will discuss these opportunities in more detail in Chapters 10 and 11.

The notion of "shared roles" may not be as daunting as it seems at first glance. The intergovernmental system under the common market regional paradigm need not be more—and may even be less—complex and troublesome than the system under which we now operate. Indeed, as we have tried to demonstrate, the current system of economic policy making has many "disabilities"[19] and may seem simpler or more efficient only because of habit.

Moreover, sharing roles does not mean that all decisions require agreement by multiple parties. Mutual or multiple vetoes represent only one form of role sharing. A system of shared roles must enable positive action as well as allow conflict.

Any arrangement for as important a matter as economic policy will involve substantial and legitimate conflict. One function of a policy apparatus, intra- or intergovernmental, is to reveal what is truly problematic and important. A second function is to manage (not eliminate) conflict and to encourage and facilitate the joining of that conflict in such a way as to produce constructive outcomes. The current system, by ignoring the regional economies, fails in both regards.

The rhetoric and politics of federalism in the 1980s and 1990s have focused chiefly on whether some policy-making functions should be allocated solely to one government or another, the sorting out tactic. In other words, the focus has been at the two extremes of the continuum regarding a government entity's involvement—no role in deciding on the one hand and an exclusive role on the other. Instead, the focus here is on the broad middle range where more than one actor has a voice, although the tenor and the authority of that voice may vary significantly across the range of situations and issues.

No one place along the continuum between the two extremes is best or right. Each situation may find its own approach. Even the midpoint, "mutual veto," need not be a dead end because the intergovernmental system is so complex that it offers circumventions and ways to begin anew.

A considerable literature and experience has accumulated that explores and develops methods of deciding—"getting to yes"[20]—among multiple actors. This work, which people have only recently begun to apply to the field of intergovernmental relations, occurs variously under such rubrics as negotiation, strategic planning, conciliation, conflict resolution, mediation, and so on. Further development of these approaches, as applied to intergovernmental relations, could considerably improve our understanding of role sharing and increase the potential for making economic policy in a system of shared roles.

This line of argument is not intended to exclude structural options but rather to increase attention to the possibilities of process. The criterion for selecting an option is that it move the situation toward effective governance and good policy for the regional commons and the common market. The metaphor of learning to dance together may be as apt as the image of renegotiating contracts.

In the end, perhaps the federalism question in its contemporary form—how to allocate roles among levels of government—should be stated differently to facilitate better answers. The issue, in the economic arena, is not just who should do what from a given list of chores and powers. The issue is who can contribute what—including newly invented strategies and tactics—to achieving the desired outcomes: effective governance and good policies for the economies held in common.

NOTES

1. Quotations can be found in Ohmae, K. (1995). *The end of the nation state: The rise of regional economics* (pp. 135, 129). New York: Free Press.

2. See, for example, *The Economist* (October 7, 1995).

3. Peterson, P. E. (1981). *City limits* (Chap. 6). Chicago: University of Chicago Press.

4. Peterson, P. E. (1981). *City limits* (p. 15). Chicago: University of Chicago Press.

5. Peterson, P. E. (1981). *City limits* (p. 16). Chicago: University of Chicago Press.

6. Peterson, P. E. (1981). *City limits* (pp. 16, 64). Chicago: University of Chicago Press.

7. Peterson, P. E. (1981). *City limits* (p. 65). Chicago: University of Chicago Press.

8. Jacobs, J. (1984). *Cities and the wealth of nations: Principles of economic life* (p. 226). New York: Random House.

9. Jacobs, J. (1984). *Cities and the wealth of nations: Principles of economic life* (p. 215). New York: Random House.

10. Jacobs, J. (1984). *Cities and the wealth of nations: Principles of economic life* (p. 214). New York: Random House.

11. Jacobs, J. (1984). *Cities and the wealth of nations: Principles of economic life* (pp. 228, 224). New York: Random House.

12. Allentuck, A. (1984, October). Jane Jacobs: Unorthodoxy from an urban visionary. *Quill and Quire*, 34.

13. For example, see Sanders, H. T., & Stone, C. N. (1987, June). Developmental politics reconsidered. *Urban Affairs Quarterly*, 22(4), 521-551(see also response by P. Peterson and rejoinder by H. T. Sanders & C. N. Stone).

14. Kuhn, T. S. (1970). *The structure of scientific revolutions* (2nd ed., pp. 52-53). Chicago: University of Chicago Press. See Chapter 1.

15. Conlan, T. (1988). *New federalism: Intergovernmental reform from Nixon to Reagan* (pp. 2, 221-224). Washington, DC: The Brookings Institution. See also Walker, D. B. (1995). *The rebirth of federalism* (Chap. 6). Chatham, NJ: Chatham House.

16. National League of Cities Institute. (1987). *Election '88: Presidential candidate questionnaire responses* (p. 50). Washington, DC: Author.

17. Rivlin, A. (1992). *Reviving the American dream: The economy, the states, and the federal government* (Chap. 7, especially p. 125). Washington, DC: The Brookings Institution.

18. Grodzins, M. (1966). In D. Elazar (Ed.), *The American system: A new view of government in the United States* (p. 8, Pt. II). Chicago: Rand McNally.

19. Lindblom, C. E. (1977). *Politics and markets: The world's political-economic systems* (pp. 323-324, 346). New York: Basic Books.

20. Fisher, R., & Ury, W. (1981). *Getting to yes: Negotiating agreement without giving in* (p. 151). Boston: Houghton Mifflin. See, for instance, Lake, R. W. (Ed.). (1987). *Resolving locational conflict*. New Brunswick, NJ: Center for Urban Policy Research; Carpenter, S. L.,

& Kennedy, W. J. C. (1988). *Managing public disputes*. San Francisco: Jossey-Bass; Kunde, J. E. (1979, November 26). Negotiating the city's future. *Nation's Cities Weekly*, pp. 25-40; Warren, C. R. (1980). *National implications of a negotiated approach to federalism* (p. 10). Paper prepared for the Roundtable on Negotiated Investment Strategy sponsored by the Charles F. Kettering Foundation and the Academy for Contemporary Problems. Regarding the actual operations of intergovernmental programs, see Peterson, P. E., Rabe, B. G., & Wong, K. K. (1986). *When federalism works*. Washington, DC: The Brookings Institution.

POLICY AND GOVERNANCE FOR THE REGIONAL ECONOMIC COMMONS

The increasingly apparent relationships between local economies and the global marketplace spotlight the importance of the regional economic commons (REC). "Competitiveness" has provided, at once, a rationale and a motivation for tending the REC.[1]

The full range of economic policy challenges that RECs face is not captured by the competitiveness label in its usual usage. If the local economic region (LER) is the fundamental spatial unit of local, national, and global economics, then citizens must look to the REC for many crucial functions. Each REC will face the task of balancing the apparent requirements of global competition with other requirements of an effectively functioning political economy.

The success of the regional political economy—the REC—in accomplishing this balancing task will largely determine the area's success, economic and otherwise. "World-class" localities, as described by Rosabeth Moss Kanter, achieve that stature; it is not an accident or a predestined status.

There remain, however, large obstacles to effective governance and good policy for local economies. In the United States, the central problem is to overcome the lack of alignment between government and economics at the local level.

There is an expanding literature about the governing and politics of metropolitan regions. More important, there is increasing experience and experimentation with varieties of governance for regions. We do not need to duplicate those stories here. A book by William Dodge, *Regional Excellence*, provides the best and most current practical information on the experiences of communities across the nation in constructing and using interlocal arrangements to address regional problems.[2]

Our purpose in this chapter is not to provide a guidebook for persons prepared to initiate action. Rather, our goal is to reframe the topic of regional governance so that such actions are more likely to succeed. The new paradigm of political economy, outlined in Chapters 6 through 8, is the basis for this reformulation.

This chapter is divided into six sections. First, we argue that (a) the standard debate about the relative virtues of fragmentation versus metropolitan government simply draws attention to the wrong questions. A new synthesis can return public discussion to a more productive focus on the fundamental issue of purpose. This synthesis is crucial for (b) addressing the urgent need to find accommodations and balances between "global" demands and "local" needs. On the basis of this reformulation, the chapter turns to (c) the options each regional citizenry might theoretically face concerning the method of dealing with the nonalignment of government and economy. Building on that choice among options, the chapter explains (d) the need for building up an adequate governance capacity in each REC. Turning to policy, we outline (e) the ways in which the kinds of problems that each REC governance arrangement will face are reframed by the regional paradigm. This includes (f) a modest proposal for federal and state roles in promoting and assisting effective governance for RECs.

BEYOND THE FRAGMENTATION DEBATE

A bipolar debate has dominated the local and regional situation for a century.[3] The central issue has been cast as the choice between mutually exclusive options regarding the structure of local government—fragmentation with no regional framework (characterized by a large number of often small governments in an area) versus replacing the multiplicity of local units with a structure of metropolitan government.

For decades, proponents of metropolitan "reform" provided the main conceptual framework for this bipolar debate. They claimed that the parochial interests of small municipalities prevented establishment of a broader, regional unit of government and that narrow interests—manifest in fragmented governments—were obstacles to the public interest. The public interest, in turn, was most often described in the terms of public administration, with proponents touting the greater efficiencies and coordination that would occur in a unified jurisdiction that encompassed the built-up area. Opponents argued that despite the administrative efficiencies that might accrue from an end to "Balkanization" (and many doubted any efficiencies would accrue), retaining government "closer to the people" was more important.

Recently, the so-called public-choice school of academic and policy scholars developed a conceptual framework that put those "fragmented," smaller units in a more favorable light. Using a marketplace analogy, they argued that many smaller units offer citizens a choice of tax, service, and governance packages; people could "vote with their feet" by moving to the jurisdiction that most fit their needs and ability to pay. The public good was thus served by the "invisible foot" of the jurisdictional marketplace.[4]

Meanwhile, local practice has often simply ignored this dichotomous debate. Decision makers developed an even more messy local situation with targeted interlocal agreements, special authorities of limited scope, and other ad hoc arrangements. From 1977 to 1987, the number of special districts alone increased from 25,962 to 29,532.[5]

In the past few years, some analysts have tried to follow local practice by formulating a discourse around "governance" rather than government.[6] This involves a double leap away from the previous debate: (a) from purely structural concerns to include (even emphasize) the process and (b) from a purely governmental focus to a broader set of civic, political, voluntary, and related capacities.

At least since the 1970s, the bipolar debate has been at a dead end. Consequently, there is a compelling need for a way out, for new formulations that incorporate the legitimate values of both poles and that create a new synthesis within which to seek better answers. The practical problem-solving approach and the governance discourse are steps in the right direction, and further steps are needed.

Defining many of these "next steps" is the regional paradigm, which creates a new synthesis that moves beyond the fragmentation debate. It refuses to choose between fragmentation and metropolitan government. It transcends the forced-choice situation and takes wisdom from both sides. It also returns the focus to the issue of purpose: In the context of noncongruent federalisms, how can we arrange governance and policy so that good decisions can accountably be made about the future of the REC?

Instead of a narrow choice between two structural options, the focus under the regional paradigm is on how to meet the operational criterion: good governance and policy for the local economy. Instead of a limited concern with better administration and efficient services, the paradigm shifts the issue to how to nurture the local economy and meet the challenge of global competitiveness. Instead of a narrow focus on government as the only answer that matters, this approach broadens the topic to incorporate the full range of governance concerns. Also, instead of a search for the one best answer, the regional paradigm's focus on purpose allows and encourages many answers.

This synthesis—created by looking at the fragmentation debate through the lens of the common market regional paradigm—is not a panacea or even an answer. It is a question. That question is the following: How can we create effective, accountable governance and make good policy for the regional economic commons?

Similarly, the synthesis does not solve problems; it identifies the right problems to be addressed. It shifts our attention away from unproductive questions to productive ones. It redefines what is problematic. Most important, it reminds us what purpose we seek to achieve and thus reestablishes a more appropriate relationship between means and ends.

This synthesis incorporates the perspectives of both poles of the previous debate. On the one hand, it accepts—indeed insists on—a regional, interlocal definition of the arena of concern: the local economy. On the other hand, it accepts the legitimacy of "local" interests.

GLOBAL AND LOCAL

There has been a tendency, in the fragmentation debate, for advocates of metropolitanism to treat local as if it meant simply "selfish" and "recalcitrant" and to regard "interests" of any but the "public" kind as somehow illegitimate. The result can be that "local interests" becomes a term of almost unmentionably bad character. This abuse of localism is not acceptable and is an obstacle to moving forward on this agenda.

In the politics of regionalism, it is important that local—even parochial and even outrageously selfish—interests be treated as legitimate (not necessarily right, just legitimate). One reason is that a "public interest" (often virtuously contrasted with local interest) is notoriously ambiguous. One person's public interest may look to another like a masked private interest. A second reason is that opponents of metropolitan government may believe that some of their important interests are best served by the fragmented situation. Dismissing localism as benighted simply does not address their specific concerns and thus will be unsuccessful in overcoming obstacles.

Advocates of regional approaches thus can take either or both of the following two approaches to opposition rooted in local or self-interest:

1. Demonstrate that the (narrow) interest will be better served in a regional venue. An example would be to argue that personal income or job stability will be maximized under an economic development program that deals with all parts of an industrial sector or "cluster" whose firms are spread over the area.
2. Provide the opponent with reasons why the special interest should be surrendered on behalf of a regionalism that will produce better outcomes for the larger number of people. This somewhat moralistic view has not and probably will not change opposition, but it can provide a rationale for the previously uncommitted (i.e., those who do not specifically identify their interest with the fragmented situation).

Both approaches are needed, but the second must not be allowed to overwhelm the first.

A dual focus on interests as well as an abstract notion of "the good" provides the basis for building coalitions that cross jurisdictional lines and contribute to the development of a real areawide politics. This is a key part of the construction of the governance infrastructure that every REC will need no matter what structure it uses. In Minneapolis, for example, state legislator Myron Orfield has undertaken a much-publicized effort to build alliances between people in the central city and those in inner-ring suburbs based on similar distress and shared interests.[7]

The changed global context makes misuse and abuse of localism even more dangerous than previously. In a real sense, more is at stake. A political approach and an acceptance of the legitimacy of localism are especially important now because some of the arguments about economic policy are being couched in terms of the need for local acquiescence to the requirements of "the global economy," as if these were beyond evaluation. In *The End of the Nation State*, Ohmae waxes a bit romantic about the benefits of free-flowing economic enterprise and the need to remove obstacles to its forward movement.[8] His zeal for spreading the benefits of global connectivity renders as wrong-headed all governmental or other reluctance to join up.

Similarly, in *World Class*, Kanter uses the term *nativist* to help explain local refusal to connect with "global" forces.[9] Equating local and nativist interests is a dangerous conflation. It delegitimizes people's completely legitimate concerns for their families' and their communities' futures. Kanter is, of course, more sensitive than that—she tries to show that connecting with "cosmopolitans" and the globalizing trends is the best way for locals to serve their interests, an exact parallel to the first approach previously recommended. The vocabulary, however, would be destructive in the hands of a less thoughtful advocate.

The local/world-class distinction is, in fact, probably too limited and limiting. Some people may care about only their place. Others may not really care about any place. The regional commons—and indeed every level of commons—needs citizens who simultaneously can care about their place and the many other places that together compose the networks on which their futures depend. These local and global balancers—not the deracinated class of international corporate professionals—are the true cosmopolitans of the twenty-first century.

Beyond the terminology, there is a more fundamental political point. The global connecting that Kanter, Ohmae, and many corporate spokespersons urge may chiefly serve only a subset of interests in the local (or national) economy. Balancing the need to connect constructively with globally active forces and the need to address the desires of other local people, households, and communities is the central challenge for every economic policymaker.

Achieving local and global balance requires specifying global arrangements accurately and then asking how various interests might be served or harmed. In the North American Free Trade Agreement debate, the addition of agreements on labor and environmental standards illustrates this strategy. (Never mind that the strategy was applied too little and too late to be effective at much more than getting the agreement accepted.) The same can be said for local hiring and subcontracting requirements added to a tax abatement for an exporting firm. These are questions that must be asked—global or "metropolitanwide" are not talismans that can ward off any opposition or tough questioning. Rather, they are generalizations that must be specified and evaluated.

Furthermore, the interests served and not served by the specific situation must be understood so that appropriate deals can be made to address needs and create effective coalitions for action. This is what it means to live in a world that has both politics and economics. Such accommodations must be made by and for each REC. This means that a wide variety of outcomes will likely occur, and people and communities can have an effective voice in determining those outcomes.

To address this central economic policy challenge of balancing global and local, each REC must get beyond the fragmentation debate and find ways to solve the central problem of the relationship of politics and economics: the fact that the two worlds do not match up—are not congruent—and are therefore very difficult to govern.

ADDRESSING NONCONGRUENCE

There are three basic options for dealing with the barrier of political and economic noncongruence. The first and second options try to eliminate it. The third option involves adapting each government's policy making or implementation or both to the fact of the noncongruence and creating systems for sharing responsibilities and benefits.

Changing Economic Boundaries

One option is to change economic boundaries to make them congruent with political boundaries. Except where natural resources or geography constitute an immovable obstacle, realigning economic boundaries would presumably be physically possible. Location decisions are discrete and could be affected by infrastructure planning, by legal requirements and penalties, and by incentives.

Indeed, this is somewhat akin to the kind of activity that state and local governments undertake when they try to lure footloose businesses or retain existing ones within their boundaries or when they establish growth management strategies. The federal government has also sought such effects in various direct and indirect spatial policies over time. Such actions are generally regarded by economists as distorting economic factors and thus as unwise.

These actions, however, are far less radical than what we are talking about here, which involves pushing all—not just selected or targeted—economic activity into a spatial pattern that is defined by a political jurisdiction. In attempting to counter powerful economic forces, such a radical effort would create extensive economic damage and waste huge amounts of current investment. It would also in all likelihood violate the commitment implied by the inclusion of property rights in the definitions of "liberty" and "the pursuit of happiness" in our commercial republic.[10] Commitment to this type of liberty is very extensive and intense in the United States.

For these two reasons—the one pertaining to economic performance, the other to economic liberty—we reject the radical option of realigning economic and political federalisms by pushing economic activity into the spaces defined by political units.

Changing Political Boundaries

Another approach to the search for congruence takes the form of trying to restructure political boundaries to be more closely aligned with economic boundaries. This would suggest that governmental entities should be constructed around the local economic region and that economic policy should be made at that level.

In general, the question here is whether political federalism can be sufficiently rearranged so that it "fits" economic federalism. No single government now encompasses one local economic region. In an attempt at resolving this problem, many proposals have been made for "metropolitan government."[11] Furthermore, it is legally and technically possible to alter state boundaries for the purpose of aligning them with metropolitan regions. This approach has the virtue of creating a government that combines the powers of state and local government.[12] In *Cities Without Suburbs*, David Rusk has argued vigorously that the more "elastic" the local government is—the more its boundaries expand to cover the settled area—the better the performance of the local economy and other measures.[13]

On balance, we cannot wholly or solely endorse the substantial effort required to create a single, large local government reigning over a wide area. Rather, we see some value in the messy multiplicity of smaller jurisdictions. The system of political jurisdictions has its own dynamics and should not be treated as a variable merely dependent on the economy.[14] The relatively unsuccessful record of metropolitan reform or consolidation suggests that we cannot realistically expect such change to occur widely anytime soon.

We are particularly concerned that the pursuit of the metropolitan government grail has tended to preclude other approaches from being tried simultaneously. It tends to induce an all-or-nothing framework that produces much more "nothing" than "all" and thus results mainly in frustration and cynicism.

As a result, leaders in a metropolitan area may have to deny explicitly any interest in structural change toward metropolitan government. Unless it is disavowed, its hovering image and the resulting polarized debate may overshadow any more complicated set of proposals. The Portland Metro, for example, is an areawide government whose experience is closely watched elsewhere. The federal transportation program, International Surface Transportation Efficiency Act, has promoted areawide planning and decision making. Councils of government exist in most places at varying levels of activity and effectiveness. Some civic and business groups have organized at the metropolitan level. Regional service delivery agencies, associations of agencies, or special authorities are increasingly in use. The media are providing, in some places, better public information on regional contexts and inter-local affairs. In most places, all of this does not add up to an effective framework for economic policy making, but it may create a good environment in which such a framework can be developed and, in some places, an institutional foundation on which to build.

Nevertheless, now is not the time to reject completely the possibility of metropolitan government. Rusk points out that a majority of metro-politan areas involve only one major county and that city and county consolidation of some kind offers an option that is relatively simple (structurally, although not politically).[15] Most important, there is in-creased interest in and practice of regional approaches.

Adapting to Noncongruence

Even if formal congruence is not achieved or even sought, every existing government's policy making can adapt to—and thus better deal with—the reality of the local economic region and the national economic system. Adaptation might involve changes, for example, in which data are collected, the spatial scope of factors to be taken into account in policy analyses, whose views are to be taken into account, the kind and extent of interlocal arrangements that are considered feasible, and so on.

In most places, adaptation is the most feasible and productive ap-proach for making use of the paradigm presented here. This is true for the following reasons:

1. Adaptation does not violate economic processes. It avoids the disadvan-tages of changing economic boundaries.
2. The adaptation approach does not require wholesale legal or constitu-tional change. It leaves the basic structure of federal, state, and local governments in place. As noted previously, the nature of the economy is not and should not be the major determinant of that structure.
3. Implementation could begin immediately and on many fronts rather than being postponed until constitutions can be amended, annexations can occur, or referenda about city and county consolidation can be held.
4. This approach allows the flexibility needed to produce further adapta-tions as the economic regions and the national economic system grow and change.

5. Substantial, although incremental, progress can be achieved by pursuing this option.
6. The cost of mistakes will be less if the approach is incremental and adaptive than if the approach is radical. People can learn from one another's experience.

Our view is that this third option—adaptation—is the most likely to provide consistently reliable basis for trying to meet the operational criterion: effective, accountable governance and good policy. Moving economic boundaries is not acceptable. In those cases in which people choose to create new governmental structures that encompass the economic territory, the performance criterion must still be met; making a "metropolitan government" is a means, not an end in itself.

CIVIC CAPACITY AND GOVERNANCE

The performance criterion for any of these three fundamental approaches to dealing with the noncongruence of the two federalisms is their capacity to nurture good governance and good economic policy for the local economy. The focus is on the need to tend the regional commons. Thus, this is not initially or primarily a matter of improving service delivery and efficiency. These things will happen as governments set out to achieve policy and other goals.

Overall, this approach creates a platform for local power. The usual objection to regional problem solving is that it requires locals to surrender control. In contrast, we argue that giving up some portion of authority over a relatively smaller arena—within the jurisdiction—may be the ticket to greater power in the wider arena—regionwide and in the common market.

Each political economy faces this governance and policy challenge distinctly. Place X may learn from Places Y and Z, but only people in X can actually make it work for X.

Thus, each REC will evolve its own distinctive culture and arrangements. How they "do" regional problem solving in the Boston area will be quite different from how they do it in Albuquerque (although both will be subject to evaluation according to the performance criterion of good governance and policy). Indeed, we believe that one advantage of the image of economics and political economy put forth here is that it encourages—indeed, requires—local variation in response to local needs and local habits.

The different facets of different REC cultures are not primarily governmental, although they are all essentially political. Crucial to the REC governance capacity are business organizations and leadership, community groups, civic organizations, processes for leadership development, and so on. Not the least important is the habit of citizens identifying themselves as citizens of the region. In some places, government or elected officials will lead, and no matter the process, they will have key roles. Governments, however, cannot function regionally in an antiregional environment. The civic culture that is supportive and the civic capacity that is enabling must be developed even as—or before—specific new arrangements might be put in place.

William Dodge has argued that approaches to regional problem solving usually have occurred in exactly the reverse order of the way they should. All too often, discussion goes to issues of mechanism (government structure) first and even only. If "a regional renaissance" is to happen, Dodge states, we must instead turn first to other key elements: making regional governance "prominent, strategic, equitable, and empowering."[16]

Regarding governments themselves, it is surely not the case that the only relevant dimension of the topic is interlocal, or between governments. Perhaps the most powerful point of leverage on the whole local political economy is within each governmental unit. In effect, the ideas of the LER and the REC and their implications for the narrow interests of each jurisdiction must be internalized by that jurisdiction.

The managers of each local government would need to understand and apply the regional economic paradigm not just to cooperate with some other jurisdiction but also to better to serve the narrow interests of their jurisdiction.

Seeing the world more clearly and correctly and then acting accordingly to serve one's own purposes is the biggest initial payoff of the proposed paradigm shift. This has been the most neglected aspect of the fragmentation-metropolitan government debate.

Of course, merely seeing is insufficient; doing is what matters. In this case, doing means nurturing the regional economic commons. Where can we start to build the capacity and culture needed to sustain a successful effort to provide governance and make good policy for the local economy? The following list[17] contains suggestions for local engagement. The list is not exhaustive, but it is intended to stimulate further thought.

1. Local governments and others, in any metropolitan area, can *inventory regional governance arrangements, experience, and capacities.* In almost all metropolitan areas, some interlocal problem solving has been and is occurring. Documenting its existence can provide precedents for further steps toward regional economic policy collaboration.

2. Local governments and other groups can undertake a variety of efforts to improve *data availability* and to describe the local economic region and work with the media and others to inform the public. Indicators of regional economic conditions, for example, reported regularly, help create basic ways of thinking and provide a foundation for public discussion.

3. *Each local government can assess the extent to which its own operations and policies, including economic development, take into account the regional context* and the needs of the local economic region. This assessment should include an analysis of the ways in which such account taking would impact the effectiveness and cost of policies and programs.

4. Local leaders can establish a regional commission to analyze and report on the *position of the regional economy in the U.S. common market and in the global economy.*

5. Local leaders can establish an *interlocal joint task force to assess the impact of poverty on fiscal and economic conditions in the area;* to examine potential connections and synergy between economic development efforts and poverty reduction efforts; and to make recommendations.

6. Local leaders can establish a task force to develop and present to the public a candid *assessment of the advantages and disadvantages of regional collaboration* and the reasons people are for or against regional collaboration in that area.

7. Local leaders can *analyze all of the economic development goals and strategies being used in the area* to identify areas of conflict, duplication, and mutual support and to assess whether they collectively meet the needs of the regional economy.

8. Local officials can *identify state and federal barriers and disincentives,* as well as incentives, for effective regional economic policy making and collaboration.

9. Local leaders could explore options for how to *identify individuals who could speak to federal or state governments* on behalf of the whole area.

Even this array of modest initiatives may seem daunting. There are plenty of good reasons for modest expectations. Despite occasional publicity in reform circles, this race will more likely go to the tortoises than to the hares. At this point, the talk about "citistates"[18] is more of a rallying cry and an ambition than an accurate description.

That ambition, nonetheless, is extremely valuable. It helps to motivate action toward the synthesis of the fragmentation debate that is achieved by looking at these issues through the lens of the regional paradigm.

Such movement toward effective governance for the REC is crucial to the economic future of the area and all its components.

POLICY AND PROGRAMS

Just as the governance debate is transformed by viewing it through the lens of two noncongruent federalisms, so too are some policy issues newly illuminated. A few examples will illustrate this point. More work will be required to light up the dark, frustration-sodden corners of current policy dead ends.

One key source of illumination is the reality and importance of the local economic region in the economic affairs of the nation. This reality is crucial to the discourse regarding competitiveness and the effects of the global economy in the United States.

In the usual form of this line of talk, there is a separate entity, an independent force that is "the global economy." It simply happens to a nation, like a change in the weather. Actions by the central government—enforcing trade agreements, subsidies for research and development, and changing national cultures of work—are seen as the appropriate and necessary responses.[19]

Through the lens proposed here, however, the global economy is the pattern of interaction among LERs and economic actors located in LERs in many different places around the world. It is the local economy, in addition to the nation-state, that is the locus of significant action.

Keen observers of transnational corporations recently have noted this shift in focus. Ohmae calls for the nation-state to "end," to get out of the way so that vigorous and ebullient region-states can do what is needed to invite in the global connections needed to thrive and grow. Kanter calls on localities to achieve "world-class" status and performance.[20] About the central importance of local economies for the competitiveness of the nation, there is no longer much doubt.

These observers, however, tend to come to this regional focus from a concern with the performance of the corporations whose success depends on their ability to navigate the sea of local economies and their linkages. This perspective provides a one-sided view of the regional commons. Much more work is needed—as we noted previously—to develop the framework for balancing the demands of transnational corporations, the requirements of local exporters and importers, the needs of domestically or locally oriented firms, and so on. Most important, how do all these relate to the future success of local households and communities and to the viability of the overall regional commons?

The performance criterion of good policy that nurtures the local economy thus highlights this issue within the competitiveness agenda: How do the apparent demands of self-appointed representatives of the global economy fit with the future viability of the commons?

Attempts to revitalize distressed inner-city areas also must be looked at through the lens of the regional paradigm. In this case, the brightest light is shed by the improved understanding of the complex internal dynamics of the local economy. In short, strategies to reduce poverty or revitalize poor communities or both can succeed only if they can take advantage of—rather than struggle against or ignore—those dynamics.

In an insightful early analysis, Mark Hughes reported on the effects, for policy, of the ways that poverty problems are defined. In *Poverty in Cities*, Hughes noted that "two main strategies" have dominated: dispersal and development. Dispersal strategies seek to "decentralize the residences of poor people" from the central city to the suburbs. Development strategies seek to "recentralize employment" from the suburbs to the city. Hughes argued that "the same forces that have fostered metropolitan decentralization (for example, transportation and communication innovations) can be exploited to connect city residents to suburban jobs." He went on to recommend a "mobility strategy" that features transportation innovations to facilitate "reverse commutes" from city to suburban job sites.[21]

Evidence has accumulated—and begun to take shape through the fog of jurisdiction-defined economic thinking—that it is the regionwide local economy that is the necessary framework for devising antipoverty strategies.

- A recent report on antipoverty strategies, done for the Annie E. Casey Foundation, places great emphasis on the metropolitan job market as the context for any successful effort.[22]
- Phyllis Furdell's study of local antipoverty economic strategies specifically calls for a regional frame for such efforts.[23]
- The guidelines for the 1994 and 1995 competition for federal "Empowerment Zone" designation contained—but did not emphasize—the local economy context for the zone strategies.[24]
- Michael Porter's essay on "The Competitive Advantage of the Inner City" is based on the premise that "inner city businesses should be positioned to compete on a regional, national and even international scale."[25]

The regional paradigm thus has important lessons and insights for the effort to devise strategies to reduce poverty. There remains much work to be done to reap the benefits of reframing this key issue.

FEDERAL AND STATE ROLES

The question of the appropriate and necessary roles for federal and state governments in promoting and assisting economic policy making for the RECs is also newly illuminated. The most important step is not, however, the providing of direct assistance. This view may be a disappointment to fans of urban assistance programs. Instead, we have two recommendations.

Like local governments, the Feds and the states must first internalize the regional paradigm and apply it. From data collection and reporting to decisions on discount rates by the Federal Reserve, the central government needs to conduct its affairs in terms of these real local economies. The same applies to each state. Chapter 11 includes further discussion of these policy possibilities.

The states hold crucial powers over the ability of local governments to create effective interlocal collaborations. These powers are political as well as legal, administrative, and financial. The degree of state influence and its current content vis-à-vis interlocal arrangements and regional problems, of course, vary. There is probably no state, however, where the situation is now sufficient.

It therefore would be useful for each state legislature to identify the impediments and encouragements to regional action currently in state law. This would be a minimal step that would provide a basis for further public discussion.

Similarly, each governor could conduct assessments, across state agencies, of the impediments and encouragements in administrative and financial practice to interlocal problem solving. One especially important item for investigation would be the ways in which state law and practice encourage interlocal tax competition for attracting businesses and thus inhibit interlocal collaboration for economic policy.

It would not be useful for either the federal or any state government simply to require some specific form of regional economic policy making. Experience with councils of governments and with community participation requirements makes clear the ingenuity with which the intentions of formalistic procedural mandates can be evaded. This is especially the case when the commitment to that intention by the federal or state government in their own affairs is negligible.

The second main recommendation here is that the federal government and the states arrange themselves internally so that they can offer to meet on a substantive agenda with representatives of any REC that can organize itself sufficiently to authorize such representation. The format for such meetings could be various. The agendas could involve any or all aspects of intergovernmental relations.

The creation, for example, of a federal (or state) team that is prepared to deal with the specifics of a regional economy and the influences, for good and ill, that the federal (or state) government now has and could have would be a significant achievement. It would also be a clear signal to serious local leaders that there may be substantial payoffs to bringing some coherence to the regional commons. Probably not many would respond initially. A few successful intergovernmental negotiations, however, would no doubt lead to more.

A modest proposal for action is a venue in which local, state, and federal governments focus seriously on the specifics of what each can contribute to economic performance in a particular local economy.

Putting it in place would constitute an important step toward applying the regional paradigm. It would encourage the new synthesis about regional governance that avoids the fragmentation and metropolitan government dead end. It would manifest an adaptation of political federalism to the reality of economic federalism. It would establish a mechanism for role sharing within which divisions of labor could more reasonably be made. Most of all, it would refocus attention on the key to all parties' success—establishing governance and making good policy for the regional economic commons. That would bring us closer to a better set of economic policies for the future of the U.S. common market in the evolving global context.

NOTES

1. See, for example, Peirce, N. R., Johnson, C., & Hall, J. S. (1993). *Citistates: How urban America can prosper in a competitive world*. Washington, DC: Seven Locks Press; Ohmae, K. (1995). *The end of the nation state: The rise of regional economics*. New York: Free Press; Kresl, P. K., & Gappert, G. (Eds.). (1995). *North American cities and the global economy: Challenges and opportunities* (Urban Affairs Annual Review, Vol. 44). Thousand Oaks, CA: Sage. See also Kanter, R. M. (1995). *World class: Thriving locally in the global economy*. New York: Simon & Schuster; National League of Cities' Advisory Council Futures Process. (1993). *Global dollars, local sense: Cities and towns in the international economy, the 1993 futures report*. Washington, DC: Author.

2. Dodge, W. (1996). *Regional excellence*. Washington, DC: National League of Cities. A very good set of analyses is provided in the following: Savitch, H., & Vogel, R. (Eds.). (1996). *Regional politics*. Thousand Oaks, CA: Sage.

3. Teaford, J. C. (1979). *City and suburb: The political fragmentation of metropolitan America, 1850-1970*. Baltimore, MD: Johns Hopkins University Press.

4. See, for example, Advisory Commission on Intergovernmental Relations. (1987, December). *The organization of local public economies* (Report No. A-109). Washington, DC: Author; Advisory Commission on Intergovernmental Relations. (1988, September). *Metropolitan organization: The St. Louis case* (Report No. M-158). Washington, DC: Author; Advisory Commission on Intergovernmental Relations. (1992, February). *Metropolitan organization: The Allegheny County case* (Report No. M-181). Washington, DC: Author; Advisory Commission on Intergovernmental Relations. (1993, October). *Metropolitan organization: Comparison of the Allegheny and St. Louis case studies* (Report No. SR-15). Washington, DC: Author. The source of the very nice "invisible foot" metaphor is Burton, R. T. (1970). The metropolitan state: A prescription for the urban crisis and the preservation of polycentrism in metropolitan society. In *Subcommittee on Urban Affairs of the Joint Economic Committee*. Washington, DC: The Urban Institute.

5. U.S. Department of Commerce, Bureau of the Census. (1987). *1987 Census of governments, government organization* (Publication No. GC87(1)-1, Vol. 1). Washington, DC: U.S. Government Printing Office.

6. See, for example, Mazey, M. E. (1991, Spring). Strategic planning. *National Civic Review, 80*(2), 216-221; Dodge, W. (1992, Fall/Winter). Strategic intercommunity governance networks ("signets" of economic competitiveness in the 1990s). *National Civic Review, 81*(4), 403-417; Wallis, A. D. (1992, Winter/Spring). New life for regionalism: Maybe. *National Civic Review, 81*(1), 19-26; Wallis, A. D. (1994, Fall/Winter). Inventing regionalism: A two-phase approach. *National Civic Review, 83*(1), 40-53; Wallis, A. D. (1994, Summer/Fall). Inventing regionalism: The first two waves. *National Civic Review, 83*(3), 290-310. See also Barnes, W. R., & Ledebur, L. C. (1994). *Local economies: The U.S. common market of local economic regions.* Washington, DC: National League of Cities; Hall, J. S. (1995). *Who will govern American metropolitan regions?* Paper prepared for the American Political Science Association Annual Meeting, Chicago.

7. Gurwitt, R. (1995, November). Man with a map. *Governing,* 96.

8. Ohmae, K. (1995). *The end of the nation state: The rise of regional economics.* New York: Free Press. See, for example, Chapter 3.

9. Kanter, R. M. (1995). *World class: Thriving locally in the global economy* (pp. 123-134). New York: Simon & Schuster.

10. See Lindblom, C. E. (1977). *Politics and markets: The world's political-economic systems.* New York: Basic Books. See Chapter 7, which briefly explains the intertwining of democracy and free markets. See discussions in Becker, C. L. (1958). *Declaration of independence: A study in the history of political ideas.* New York: Random House; Elkin, S. L. (1987). *City and regime in the American republic.* Chicago: University of Chicago Press.

11. Coleman, W. G. (1975). *Cities, suburbs & states: Governing and financing urban America.* New York: Free Press. This reference catalogs a series of such proposals. Regarding "metro government," see Canty, D. (1972, March/April). Metropolity. *City,* 29-44.

12. Richard P. Burton recommends city and state constitutional management in Burton, R. P. (1970). The metropolitan state: A prescription for the urban crisis and the preservation of polycentrism in metropolitan society. In *Subcommittee on Urban Affairs of the Joint Economic Committee.* Washington, DC: The Urban Institute. Jane Jacobs calls for a similar approach in Jacobs, J. (1984). *Cities and the wealth of nations: Principles of economic life* (Chap. 13). New York: Random House. Recently defeated proposals for changes in Rotterdam, The Netherlands, had similar interesting characteristics.

13. Rusk, D. (1993). *Cities without suburbs.* Washington, DC: The Woodrow Wilson Center Press.

14. On the autonomous state, see Gurr, T. R., & King, D. S. (1987). *The state and the city.* Chicago: University of Chicago Press; Jones, B. (1990, March). An uneasy certitude in urban political economy. *Urban Affairs Quarterly, 25*(3), 527-529.

15. Rusk, D. (1993). *Cities without suburbs* (pp. 91-98). Washington, DC: The Woodrow Wilson Center Press.

16. Dodge, W. (1996). *Regional excellence.* Washington, DC: National League of Cities.

17. This list is reprinted with permission from Barnes, W. R., & Ledebur, L. C. (1994). *Local economies: The U.S. common market of local economic regions.* Washington, DC: National League of Cities.

18. Mahtesian, C. (1995, September). Regionalism I: Peirce's prescriptions. *Governing,* 24-28.

19. The newsletter of the Council on Competitiveness (Washington, DC) is especially interesting in this regard.

20. Ohmae, K. (1995). *The end of the nation state: The rise of regional economics.* New York: Free Press; Kanter, R. M. (1995). *World class: Thriving locally in the global economy.* New York: Simon & Schuster.

21. Hughes, M. A. (1989, February). *Poverty in cities.* Washington, DC: National League of Cities. Quotations are from p. 16. See also Hughes, M. A. (1989). *Fighting poverty in cities: Transportation programs as bridges to opportunity.* Washington, DC: National League of Cities.

22. Dewar, T., & Scheie, D. (1995). *Promoting job opportunities.* Baltimore, MD: Annie E. Casey Foundation.

23. Furdell, P. (1995). *Paths to economic opportunity.* Washington, DC: National League of Cities.

24. U.S. Department of Housing and Urban Development. (1994). *Building communities: Together.* Washington, DC: Government Printing Office.

25. Porter, M. (1995, May/June). The competitive advantage of the inner city. *Harvard Business Review, 55.*

THE UNITED STATES COMMON MARKET
Policy and Governance

Governance and policy for the whole common market require participation by local, state, and federal governments as well as by representatives from the regional economic commons (RECs). The common market is not just another name for the federal government in its economic policy mode.

This complex intergovernmental situation, engendered by looking at the economic world through the regional paradigm, is a horizontal federalism. It flattens the pyramid of vertical federalism. Creating governance and making policy for the common market require understanding and addressing the full scope of interaction among the three systems described at the end of Chapter 8: the political economies of the regions, the common market, and the global systems. This situation raises the following major challenges for all participants, including states, local governments, the federal government, and regional commons spokespersons:

- It requires each government and REC to develop and hold truly "national" and truly "local" perspectives simultaneously.
- It does not assume that national equals federal. (National becomes an achieved rather than an ascribed status; national is as national does.)[1]
- It also redefines national as a complex web of systemwide interactions whose outside borders are ambiguous.

This amounts to a reformulation of the participants list for what would be called "national economic policy making" using the nationalist paradigm. In that framework, the federal government has pretty much exclusive responsibility and authority in the economic arena. Under the rubric of governance and policy making for the common market, however, economic policy must be shared because the economy is shared.

Under the nationalist allocation of roles, there is no need to search for a way to bring the players together; the federal government is appropriately ready and seeking. Under the regional framework, there is no such easy selection of participants.

There is clearly a need for an effective venue in which stakeholders and policymakers can come together. No such venue now exists.

The development of effective governance and policy will not, of course, be limited to one prominent venue. Adapting political federalism to address the needs of the common market will involve a variety of efforts at many points in the intergovernmental system.

In this context, the roles and policy arenas of the federal, state, and local governments will change. The dominant mode of intergovernmental relations around economic issues must be—as we argued in Chapter 9—one of shared roles and a greater clarity of purpose: to enhance the condition and performance of the common market of regional economic commons.

This chapter will provide a framework within which further development and application of the regional paradigm can occur. It begins with some ideas about how the federal and state governments can proceed along these lines even without any new intergovernmental forum. The chapter turns next to the policy agenda for nurturing the economic system and the ways in which problems may be reformulated when seen through the regional paradigm. Finally, it returns to the question of governance and especially the problem of the current absence of a forum where the stakeholders can meet and engage one another on common market policy and politics.

FEDERAL AND STATE ACTIONS

The recommendation for shared roles and multiple players on economic issues is not the same as the call for "devolution" currently heard in Washington and in some governors' mansions. The federal government has hugely important roles to play; it is just that they are not exclusive anymore.

The power of devolution politics, however, does indicate that decentering the Feds may not be as difficult as it might have been in earlier heydays of nationalist economics.

Carrying forward the implications of the regional paradigm, however, is not necessarily a matter of high politics. Federal and state action on these matters need not—and ought not—wait for a new piece of grandly omnibus legislation. Small and intermediate steps, steps that are well within the purview and powers of federal and state officials, will be constructive contributions.

The key guideline is the need to internalize the regional paradigm into the normal business of these governments. That is where a new way of seeing the economic world will be most useful and will have the most immediate effect.

The following are some such steps that might be initiated by the federal government.[2] The list aims to illustrate the point that small, but important, steps can be taken. It may also have the effect of stimulating further thought by readers:

1. The Office of Management and Budget (OMB) can coordinate the relevant agencies to *assess the quality and availability of federal data* needed to support research to develop indicators of local economies' capacity and performance. OMB would make recommendations about needed im-

provements and about how to make better use of what is now available. The effectiveness of many other initiatives depends on adequate data.

2. *The Federal Reserve System* can develop more systematic reporting and analysis on local economic regions for its deliberations and for the *Beige Book* (the periodic report on conditions in each Federal Reserve region).

3. The president can establish several demonstration *Intergovernmental Labor Market Task Forces*, each focused on one local economic region, to identify and seek to reduce or remove impediments to efficient and effective functioning of the local labor market. Each task force would include federal, state, and local representatives and could be convened by the Secretary of Labor.

4. The Executive and the Congress can ensure that regional economic commons' voices are heard in the next phases of the debate about the national information infrastructure (NII)—*the "information superhighway."* This significant infrastructure will be crucial to the economic growth of local economies. The White House can, for example, convene an NII forum that uses a specific metropolitan area as an example to explore the local dimensions of the NII, including capacity to participate, impacts of innovations, barriers to access, the balance of public and private interest, the potential for economic growth, and so on.

5. The Council of Economic Advisors can include a chapter in the *Economic Report of the President* on local economies.

6. The president can direct his staff to coordinate an analysis, in every department and agency, of *the impacts of existing programs* on regional cooperation, the performance of local economic regions, and the reduction of intraregional disparities.

7. *Congress can hold hearings on the relationships of investment strategies to common market regionalism.* One focus could be, for example, to reexamine the "Metropolitan Trust Fund Initiative," submitted to the Senate's Urban Revitalization Task Force in May 1992 by Tim Honey with John Tayer. It proposes (a) federal 2-year matching grants to metropolitan economic regions for formulating long-term economic development plans and (b) federal support for implementation of those plans.

8. The president can establish a mechanism to *develop and experiment with targeted antirecession tools* that could be used in the future. The triggers for using those tools would be some combination of national average indicators and local economic region (LERs) indicators. The targets for the tools would be downturning LERs and particularly vulnerable areas and groups within them. (Improved and timely data will be important for this.)

9. The *Advisory Commission on Intergovernmental Relations* (ACIR) can examine and report on the implications of the common market regionalism paradigm for federalism.[3]

10. The Secretary of Housing and Urban Development can establish an *Intergovernmental Local Economies Advisory Group,* in conjunction with which the secretary could conduct an ongoing consultation and discussion with local leaders about the importance of local economies and their interaction with housing markets and with housing and development policies.

A somewhat analogous list could be developed for each state.

PROBLEMS AND POLICIES

The paradigm shift described in previous chapters creates or reveals the need for accommodation by political federalism. That shift also changes our understanding of the problems that policy for the common market must try to solve. It opens new areas for potential solutions and new resources and capacities for implementing them.

The most fundamental policy implication is that the locus of concern in most policy issues must go first to the LERs. The telling economic indicators and the crucial economic results are about the condition and performance of those basic building blocks.

Thus, for example, the loud concerns about the "competitiveness" of the United States' economy suffer from a misplaced concretion. It is the performance of the local economies and the capacities of the RECs that will have most to do with U.S. standing in the global patterns of economic activity. It is there, not in the national averages, that the important things will happen on everything from the quality of schools and workforce development to the adequacy of infrastructure systems and the efficiency of industrial clusters.

Even the way we develop trade policy should be reconsidered. Rather than giving exclusive attention to industrial sectors, analysis should also look at the ways LERs are or will be affected by flows across national borders. Currently, there is now no such analysis and no part of the process in which economic places—that is, local economies and their commons compared to, for example, governments and corporations—can have a voice.

Giving the RECs a voice in the development of trade policy may result in some different outcomes. One guess would be that there would be continued emphasis on opening up trade flows and that there would be more emphasis on finding ways to provide trade-offs for people and communities that are negatively affected by such flows, at least in the short run.

Policies directed at evening out the business cycle would also need to shift attention from national averages to the conditions of LERs. Each local economy has its own cycle and each would need to be monitored. Clearly, forward and backward linkages transmit changes from one to others, so downturns will exhibit some patterns across the nation.

The nature of those patterns is not to be assumed; it is to be determined by empirical observation. That means that the quality and timeliness of data for the LERs should be significantly upgraded.

Efforts to mitigate those downturns and to moderate upswings should also be focused on the LERs and the patterns among them. Whether Keynesians or monetarists, federal economic policymakers have focused on the national average indicators and on mitigating national average business cycle patterns. This is not sufficient. There is much room for experimentation by the Federal Reserve and by the Congress in developing tools that can achieve these LER-targeted results.[4]

The application of the common market regional paradigm to policy should proceed into all relevant policy arenas. Both general and very specific reorientations from the currently dominant framework will be needed. This will prove very difficult, if only because the assumptions of the prevailing nationalist framework are deeply embedded. Topics that need reexamination include, for example, infrastructure, capital formation, financial regulation, workforce development, education, small business development, and defense conversion. The list includes every and any issue on the economic agenda.

The initial agenda for common market governance should be constructed, however, with a concern for the aspects of the system itself rather than as a list of specific policy issues (which will, of course, quickly follow). That agenda will include at least the following five major topics:

- the linkages among the LERs (e.g., transportation, communications, and networks for policy innovation);

- the adequate performance of all LERs in the economic system;
- the economic system's common elements (e.g., currency, monetary policy, and regulatory frameworks);
- economic relations between the common market (and its constituent LERs) and other common markets and nation-states; and
- the governance process itself.

This system agenda is important because it focuses on the way the economy and the political economy work in the world of common market regionalism. The normal list of policy topics—such as in the previous paragraph—is imbued with the nationalist or statist paradigm; starting with them will make the journey even more difficult. In the initial stages, however, the two agendas will both be present to facilitate a transition and to increase the early appeal of the effort.

TOWARD GOVERNING
THE COMMON MARKET

The figures in Chapters 6, 7, and 8 show economic systems in which all the parts are different from one another and all are importantly linked to other parts. This interrelatedness has implications for political federalism and for governance of the two political economies: the regional economic commons and the common market.

The following two main implications have already been identified, in somewhat different form:

1. Each existing government must at least minimally internalize the regional paradigm so that all participants are looking, so to speak, at the same object (the local economy in the case of the REC; the economic system in the case of the common market).
2. Each government must acknowledge the legitimacy of the interests of the others and must be able to take those into account in decision making for economic policy.

The second has a familiar ring to it. It is the mode of a legislative or associational body, in which only the interaction of the parts produces the whole, the "public interest." This is the image of Madisonian republicanism, in which "factions" are everywhere but are herded by the system into paths that lead to the better good of all.[5] This idea and way of proceeding is not new to Americans active in government and politics.

The new question is where can this confluence of interests occur in the new political economy described here? Specifically, governance is required for each REC and for the whole common market. In neither case are the three-tiered "levels" of governmental federalism directly useful. As for the RECs, Chapter 10 called for people and institutions in each area to put in place whatever means they can to meet the criterion of effective governance and policy for the local economy. Some important contributions were noted that must be made by the federal government and especially by the states.

At the geoeconomic level of the common market, additional intergovernmental dimensions open up. Most significantly, the distinction between "national" and "local" becomes harder to maintain. (The federalism pyramid is flattened.) In effect, the federal government must act nationally and think regionally, and local (and state) governments must act locally and think regionally. Each takes on some of the mentality and perspective of the other by being engaged in this immensely interactive—though often conflictual and inefficient—relationship.

Thus, as at the regional commons, the issues are most importantly governance and purposes rather than structure. As at the regional commons, there is no structural fix for the question of "who will speak for the commons?"

No existing government, including the federal government, can adequately address this question. The central issue is what each government can contribute to policy making for the common market, which does not belong solely to any of them because political and economic federalisms are not congruent. Furthermore, no existing government can speak for any individual REC nor for the systems of RECs, but the RECs must have voice in common market governance.

Therefore, new arrangements are needed to facilitate governance and policy making for the common market. In the longer term, a variety of options are possible and a range of institutional capacities may be useful. In the near term, a forum for the common market is urgently needed.

The forum is an adaptive, not a radical, strategy. The analysis of basic options for achieving governance that overcomes the noncongruence of the two federalisms (presented in Chapter 10) also applies here. Adaptation is probably the only reliable basis for constructing this governance capacity. New forms and processes of "getting to yes" can be applied. Nurturing the economic system, the operational criterion, is more important than any particular mechanism.

The call for a forum for the common market is not a call for some new government. It is a recommendation that governments come together to address the needs of the economies and the systems of economies for which they share responsibility and on whose performances their fates depend. Some such arrangement for a convocation of stakeholders is crucial to the economic and political future of the United States.

The focus of the forum would be on economic conditions and issues in the LERs, the linkage systems, and thus throughout the common market. The purposes of the forum would be to move toward some triangulation of local, state, and federal government policies regarding this focus and to develop an ongoing governance capacity for the common market. The initial agenda would seek some balance between near-term policy results and the next steps toward establishing the forum or its successor.

Voices from and for the RECs should be heard in the forum. The opportunity to participate will create an incentive for stakeholders in each region to come to some agreement about representation and accountability. That may not be an easy or attractive process in all cases.

Similarly—but even more difficult—the forum creates a voice for the common market. No existing entity supplies that voice; the federal government is not the common market. That voice can be expected to be multitudinous and discordant but to move—over time—to some patterns that persist.

The choice of a first step toward the forum would be pragmatic. Convening an event would provide drama and specificity. The president could call together the forum (President Clinton did something like this in December 1992, but the agenda and participants list reflected the conventional nationalist paradigm), or a group of REC representatives could issue a call. The agenda for this event, as noted, would be a combination of policy concerns and steps toward governance capacity.

Another step could be a report on the state of the common market. This report would provide information and frame aspects of a public discourse. It could be issued periodically, with some portions issued more frequently than others, depending on the availability of data. A foundation or a set of stakeholders would have to provide funds.

What would the forum look like over time? For some, the experience of the European Community (EC) and other examples from outside the United States regarding governance and policy may be helpful in thinking about the forum or other mechanisms. The EC is a complex entity in which the economies and the governments are not congruent. Unlike the United States, it did not start with the overarching "national" government in place to create a common currency and eliminate internal trade barriers. That is surely the end of any analogizing that might tempt U.S. observers. The EC's evolving "regional policy" agenda may be of particular interest. Although the experience will be worth analyzing, the requirements of the common market as conceived here are different from the nationalist basis of thought and practice in the EC. In the end, the forum will have to be invented, not copied.

Besides the forum, the interlinked nature of the economic system will require and probably engender other new structures and collaborative processes. Federal, state, and local governments, along with the REC governance representatives, will have to combine in various ways to address such situations as the following:

- Individual LERs that cross state lines
- Individual LERs that cross national boundaries
- Sets of LERs that are contiguous
- Sets of LERs that are closely linked (e.g., because they are dominated by a similar industrial sector or because one is dominated by suppliers of the industry that dominates the other) but are not contiguous
- U.S. LERs that are closely linked but not contiguous with one or more LERs abroad
- Sets of RECs that want to make bilateral or multilateral agreements among themselves

Some such adaptations to the regional nature of the U.S. and global economies are already under way. On the international trade front, for example, efforts from the Seattle Trade Alliance and "Cascadia" to the

Carolinas Partnership seek to organize RECs to compete and to use the capacities of all levels of government to enhance the performance of the local economy.[6]

These practical efforts simply ignore the nationalist paradigm. Within that paradigm, they are anomalies that arise because it is applied to a world it no longer adequately describes or explains.

We are well beyond the time when energy must be wasted adjusting the nationalist paradigm to anomalies or, worse yet, adjusting those "anomalies" to the outmoded paradigm. The regional paradigm provides a better description and explanation, one that can serve as the rationale and guide for constructive action.

All governments will better serve their own interests and will help nurture the several commons by following the strategies proposed here: internalizing and applying the regional framework and developing governance arrangements—including the forum for the common market—adequate to provide good policy.

NOTES

1. Arguing that local governments are not necessarily "closer to the people" than states or the federal government, Grodzins wrote that "local is as local does." We have analogized his bon mot. Grodzins, M. (1966). In D. Elazar (Ed.), *The American system: A new view of government in the United States* (Chap. 7). Chicago: Rand McNally.

2. This list is reprinted with permission in a slightly revised form from Barnes, W. R., & Ledebur, L. C. (1994). *Local economies: The U.S. common market of local economic regions.* Washington, DC: National League of Cities.

3. At this writing, ACIR is headed for a zero budget and limbo in FY 1997. This is a terrible idea. The ideas we propose in this book make it all the more important that fresh attention to the conditions of American political federalism be continued.

4. A recent study, for example, asked, "Does monetary policy have differential regional effects?" The answer, according to the study, is yes. Carlino, G. A., & DeFina, R. H. (1996, March/April). *Business review* (pp. 17-27). Philadelphia: Federal Reserve Bank of Philadelphia. On the effectiveness of subnational targeting of monetary policy, see two interesting discussion papers from Canada: Miller, F. C. (1979). *The feasibility of regionally differentiated fiscal policies* (No. 1979-6); Miller, F. C., & Wallace, D. J. (1982). *The feasibility of regionally differentiated fiscal policies: Some further results* (No. 1982-7). (Miller was at the University of Guelph and Wallace at Statistics Canada).

5. Madison, J. (1961). Federalist Number 10. In *Alexander Hamilton, James Madison, and John Jay, the Federalist Papers* (pp. 77-84). New York: New American Library; Madison, J. (1961). Federalist Number 51. In *Alexander Hamilton, James Madison, and John Jay, the Federalist Papers* (pp. 320-325). New York: New American Library.

6. See National League of Cities' Advisory Council Futures Process. (1993). *Global dollars, local sense: Cities and towns in the international economy, the 1993 futures report.* Washington, DC: Author. See also Kresl, P. K., & Gappert, G. (Eds.). (1995). *North American cities and the global economy* [Urban Affairs Annual Review, Vol. 44]. Thousand Oaks, CA: Sage; Brooks, J. (Ed.). (1995). *Local officials guide to leading cities in the global economy.* Washington, DC: National League of Cities.

AFTERWORD
The Challenge and the Opportunity

The preceding chapters have argued the following points for a better framework for thinking about and acting on economics and political economy:

- Local economic regions are the real economies—the basic, functional economic units.
- These metropolitan-centered, regional economies are the building blocks of the three-tiered economic system—regional, national, and global. No part of this system is autonomous; linkages and interdependence pervade the networks.
- Political and economic boundaries are not congruent. The conventional nationalist assumption that they are is incorrect and results in flawed policy.
- Political economies thus face the dual challenges of nurturing the economic commons for each of the three tiers and balancing the demands of each against the others.

173

These and related ideas no longer can be accommodated within the dominant way of thinking about "the national economy." The prevailing nationalist economic paradigm does not adequately describe the real world and, as a result, does not provide a reliable and useful foundation for policy making. Therefore, we recommend a shift from this fatigued framework to the common market regional paradigm.

REFRAMING THE ISSUES

These ideas have consequences. Changing the lens through which we see the world leads us to reframe—in ways both large and small—how we understand the problems and the issues we face.

Under the regional paradigm, the "global economy" becomes less a rarefied entity and more a set of patterns of worldwide interaction among local economies. Imperatives from observers such as Kenichi Ohmae and Rosabeth Moss Kanter that local leaders "invite in" or connect with the global economy may seem to envision some free-floating economic oversoul. Instead, they can best be understood as recommendations to seek out constructive engagement with other local economies, whether directly, through transnational financial institutions or multinational corporations, or through new arrangements among nations. Local to global is abstract and unreal; local to local is the connection that must be made.

The new lens of the regional paradigm also helps move things beyond fruitless either-or debates. We have argued for putting aside the nationalist, or statist, perspective. We have also, however, rejected simple economism. The complex political economy should be seen whole if effective policy is to be developed and implemented.

Thus, the nation-state remains important in the new context. Its demise is not in sight; the nation-state must adapt, not abdicate.

Similarly, the regional paradigm reveals as pointless the debate between so-called free trade and so-called protectionism. There is no use denying the overall benefits of fewer restrictions on economic interaction. Also, there is no use denying that some people, firms, and communities will be harmed, not helped, by such interactions. Moreover, it is clear that rules are needed and that not all "harm" can or should be prevented. The challenge is to accept both facts—to embrace the paradox of "creative destruction," in Schumpeter's[1] phrase—and to develop approaches that minimize and indemnify substantial harm while obtaining for ourselves the benefits of more open economic interactions. This applies to the national common market as well as to the regional economic commons.

Chapters 9 and 10 argue for a synthesis of the all too often bipolar debate on governmental fragmentation versus metropolitan government. There is much that can be done about capacity building for regional problem solving. Only some of it pertains to government structure. The regional lens allows us to see beyond the thesis and antithesis cul-de-sac to a more open road of constructive action.

Regarding the vertical dimension of political federalism, there is a similarly urgent need to get beyond the "sorting out" debate. This scholastic exercise among governmentalists is driven more by a compulsion to escape budget quandaries than by a search for ways in which government can meet its responsibilities. It pretends that "roles" are exclusive and ignores the obvious, albeit messy, reality that shared roles are necessary to address problems in a complex world.

Instead of a statist focus on which government entity will do what, we recommend that the initial focus—in the economic policy arena—be on the desired outcome: nurturing the regional economic commons and the common market. Then the issues of intergovernmental relations have an operational criterion by which they can be evaluated.

These regional economic commons, it must be noted, are not reserved to big cities alone. Kanter's "world-class"[2] capacity is a goal for all regional commons.

The new lens helps clarify that the narrow basis for much local economic development today must be broadened. Reducing poverty and inequality and alleviating the economic insecurity of the working class are economic issues that are part of, not separate from, the overall economic policy challenge that often goes under the misleading rubric of "competitiveness." Governance capacity and workforce quality are also crucial. A narrow focus on reducing firm costs is not at all sufficient.

IMPLICATIONS FOR GOVERNANCE

In addition to calling for a change of paradigms and for a careful reframing of issues through the new lens of the common market regional paradigm, we have made the following four main recommendations for local, state, and federal government action:

1. Local governments and other stakeholders should use the regionalism framework for their own policy making and get together to form a regional economic commons (REC) to tend the local economy (the local economic region).
2. Each state should use the regionalism framework for its policy making and should arrange itself to meet with and address the needs of each REC that contains area within that state.
3. The federal government should use the regionalism framework for its policy making and should arrange itself to meet with and address the needs of each REC that contains area within the borders of the United States.
4. A forum for the common market should convene as a first step toward governance and policy making for the U.S. common market of local economies.

Taking these steps will require substantial adaptation by all governments—federal, state, and local. Internalizing the implications of the common market regional paradigm will also reframe the interests of households, firms, and communities. These governance recommendations will not put an end to conflict. Rather, they will shift conflict as well as coalition making into more productive channels. One welcome contribution may be to shed new light on struggles over key economic issues frequently fought out in city trenches labeled zoning, land use, bus service, licensing, and so on.

Another would be to bring the fresh air of local reality into abstract federal and state policy development. (Again, this is not an argument for "devolving" functions; it is a mandate for better federal and state policy making and capacity building.)

A NEW ARCHITECTURE

Our call for regional commons and for a common market mechanism outlines an architecture, a framework, for the political economy we have described. Each commons keeps others in bounds, creates an identity for shared interests, and also provides for the play of divergent or narrow interests. This is federalism, not feudalism. The linkages, tensions, and conflicts within each REC, among the RECs individually, and between the RECs and the common market collectively are the posts and lintels for the development of a balanced civics. Multiple tiers and multiple commons mean that citizens have multiple identities and loyalties. In that multiplicity are the dissonances that help people hear one another and overcome separatism. The house of the new political economy will be stick-built, not modular, and it will be as dependent on countervailing tensions among interests as on firmly nailed joining for the common good.

The capacity of a commons at each tier to speak into a commons at the other tiers is thus a key aspect of this institutional framework. The lack of such capacity at the global tier is a great deficiency and danger; transnational corporations will not meet the need. Whatever evolves, it must go beyond the nation-state basis (of, e.g., the United Nations) and it must create an effective discourse with regional and common market tiers.

Much is at stake in these ideas and proposals. A deeply flawed economic framework produces faulty policy, which, in turn, has ill effects on economic performance. Better ways of understanding the economic world would reduce the likelihood of that sad sequence persisting and could result in better policy. Under the common market economic paradigm, for example, the focus for competitiveness would be the healthy performance of local economies—not abstract national averages—and the capacity of the regional economic commons to organize itself for action. Greater understanding of the highly interactive nature of the economic system—and of the fact that problems in any one place harm the entire system's economic performance—would result in a search for strategies to reduce poverty and distress, strategies that would contribute to better outcomes for the whole system. Anti-inflation and antirecession efforts would be targeted to the local economies that need them. Workforce development initiatives would focus on building capacity for connections and lifelong learning throughout the regional economy.

Equally important, there is much at stake for politics and government. First, the regional paradigm and the kinds of governance approaches we recommend make the central political issue of "the economy" more tangible, more understandable, less remote, and less abstract. It is bad for democracy that the policy arena for "economics" is so far removed from the lives of citizens that leaders can report—with apparent sincerity and without fear of reproach—that the economy is booming and yet most people are not doing very well. The local economic region brings economics down to earth. It shows how what happens locally is connected to things that happen elsewhere. In short, politics can thrive where the issues can be understood.

It should be clear that the topic here is not just government programs. The political economy of common market regionalism is already evolving a corollary civics and social capital. This evolution is uneven. The world class of symbolic analysts has its Internet and its multinationals; locally rooted development interests have moved to a regional scale. "No-roots" and "all-roots" (the dangers of "McWorld and Jihad" that Benjamin Barber has warned about),[3] however, are not sufficient to sustain a civics for the new political economy. The recognition of the regional framework is a challenge for all citizens and one that must be met if an effective politics is going to thrive.

Second, the new paradigm changes what we "see"—and, especially, what we see as important and as problematic. Thus, it opens the possibility of redefinitions of shared and hostile interests and of new alliances and coalitions. Coalitions grounded in good understanding of the real world, rather than in abstractions, will create a more realistic, more robust politics.

Finally, better policy will produce better results. In turn, that will make politics less frustrated, especially by diminishing the perception that the economy cannot really be dealt with—that it is beyond us to influence it in the directions we want it to go. Efficacy is also important to a thriving politics.

FURTHER RESEARCH

It is important that both research and practical experimentation proceed side by side on these issues. Ideas matter, but they matter most when someone does something about or because of them. The key is that thinking and doing go forward in ways that interact so that each activity can build from the other.

It is better, as the dictum goes, to describe and measure the right thing crudely than the wrong thing precisely. Better still is to measure the right thing precisely. Reliable numbers about economic conditions in small areas—including areas that are not defined by governmental jurisdictions—will be crucial to further development of the regional paradigm. Data help create and define social science reality. Current data reflect and reproduce the reality of the nationalist paradigm.

Recent publication of export data by metropolitan areas provides an example. Previous data informed us only about U.S. totals and about points of debarkation, thus treating the nation as a homogeneous unit. The Census Bureau has undertaken a commendable effort to improve these numbers so that the analyst can compare, for example, the export sectors of various metropolitan areas. These data will be most useful if, for example, they allow description by economic clusters rather than by (homogeneous) local government jurisdictions.

There are many obstacles in improving the data: conceptual clarity, logistics, privacy issues, interagency competition, and diminishing funding. One encouraging factor may be that proposals for some kind of interagency coordination or even consolidation are receiving more serious consideration. Alone, however, this might result in a rigid bureaucratic fortress committed to defense of the conventional paradigm. A second promising development is the potential for local, state, and federal cooperation in this area, reflected in part in the work of the Advisory Committee for the Year 2000 Census. A more drastic, longer-term approach would be to shift economic data responsibility to an agency of some successor to the forum for the common market.

A second area for further research is to describe the local economic regions (LERs) and their linkages in specific terms. Boundaries, internal dynamics, and linkages should be detailed for each LER and for the national and global systems. These are immense, complex tasks that would be continual, not one-time. They depend eventually on data development and, in turn, would help shape new data. The various economic models that are now in use for specific metro regions are a beginning. More of these are needed, and the question of how they interface must be addressed. Similarly, more case studies and comparative analyses of how government and politics work (and do not work) in the RECs will be important. The reformist stories by Peirce and colleagues were a very useful step; the effort by Savitch and Vogel and contributors is a further step.[4] Studies of the RECs will constitute additional dimensions of work: case studies of various linkage mechanisms, clusters of integrated LERs, common market kinds of functions, and so on.

One especially urgent topic is the relationship between distressed LERs and others and between distressed parts of an LER and the rest of it. The issues posed in Chapter 7 about the ways in which disparities—especially poverty—affect growth and vice versa need attention soon lest rapid restructuring in the interests of the nonpoor and the nonlocal render those questions effectively moot, short of outright redistributional conflict.

A reasonable test of a proposed "paradigm shift" is the richness of the research agenda it can engender as well as the utility of the practical steps it fosters. The previously mentioned items just begin the list of areas that this perspective opens or reopens for investigation. Not the least among these is the further elaboration of the lens—the architecture, the paradigm itself, and the framework, as we have variously described it.

THE CHALLENGE AND THE OPPORTUNITY

In the thinking we have offered, we have built on the work of others, not unquestioningly but with appreciation. A measure of success will be that still others will now build on our work.

Changing the way we think about economics and government is no small challenge. Tendrils of habit and the kudzu of vested interests constrain this effort. The task is to step outside that grove of conventional thought and to look again at the whole forest of political economy.

Then we can develop and test the utility of approaches that better reflect how the political economy works. This is the challenge and the opportunity.

NOTES

1. Schumpeter, J. (1934). *Business cycles*. New York: McGraw-Hill.
2. Kanter, R. M. (1995). *World class: Thriving locally in the global economy*. New York: Simon & Schuster.
3. Barber, B. R. (1995). *Jihad vs. McWorld*. New York: Times Books/Random House.
4. Peirce, N. R., Johnson, C., & Hall, J. S. (1993). *Citistates: How urban America can prosper in a competitive world*. Washington, DC: Seven Locks Press; Savitch, H., & Vogel, R. (Eds.). (1996). *Regional politics*. Thousand Oaks, CA: Sage.

INDEX

183